PRACTICAL BLACKSMITHING

A COLLECTION OF ARTICLES CONTRIBUTED AT DIFFERENT TIMES BY SKILLED WORKMEN TO THE COLUMNS OF "THE BLACKSMITH AND WHEELWRIGHT" AND COVERING NEARLY THE WHOLE RANGE OF BLACKSMITHING FROM THE SIMPLEST JOB OF WORK TO SOME OF THE MOST COMPLEX FORGINGS

VOLUME III

Compiled and edited by
M. T. RICHARDSON

ILLUSTRATED

Published by Left of Brain Books

Copyright © 2022 Left of Brain Books

ISBN 978-1-396-32140-5

First Edition

All rights reserved. No part of this publication may be reproduced, distributed, or transmitted in any form or by any means, including photocopying, recording, or other electronic or mechanical methods, without the prior written permission of the publisher, except in the case of brief quotations embodied in critical reviews and certain other noncommercial uses permitted by copyright law. Left of Brain Books is a division of Left of Brain Onboarding Pty Ltd.

Table of Contents

PREFACE.	1
CHAPTER I. Blacksmith Tools.	2
Their Preservation.	2
Bench Tools.	4
Blacksmith Tongs.	8
How to Make a Pair of Common Tongs.	8
Tools for Farrier Work.	9
A Tool for Holding Plow Bolts.	19
Tongs for Holding Plow Points.	19
A Tool for Holding Plow Bolts.	20
A Tool for Holding Slip-Shear Plows in Sharpening.	20
Making a Plow Bolt Clamp.	22
Tongs for Holding Plow Bolts.	23
How to Point a Plow.	23
Hints for Plow Work.	24
A Tool for Holding Plowshares.	26
CHAPTER II. Wrenches.	28
Forging Wrenches.	28
Making Wrenches.	29
Curve for an S Wrench.	30
Another Method of Making Wrenches.	32
An Adjustable Wrench.	33
Making an Adjustable Wrench.	35
Forging a Bolt, a Nut and a Wrench.	35
CHAPTER III. Welding, Brazing, Soldering.	38
Melting Point of Metals.	38

Solders.	39
Fusible Compounds.	40
Fluxes for Soldering or Welding.	40
Theory of Welding.	40
Welds and Welding.	44
Welding Iron and Steel.	53
Points About Welding.	53
Welding Cast Steel Forks.	54
Welding Steel.	55
Welding Steel Tires.	55
Welding Tires.	55
Do Not Burn Your Tires in Welding.	56
Welding Axles.	57
Welding Cast Iron.	57
Welding Malleable Iron.	58
Welding Malleable Cast Iron Plates.	58
Welding Cast and Wrought Iron.	58
Welding Steel to a Cast Iron Plow Point.	58
Welding Plow Lays to Landsides.	59
To Weld Cast Steel.	59
To Weld Steel Plate to Iron Plate.	59
A Practical Method of Welding Broken Spring Plates.	60
Welding Buggy Springs.	61
Welding Springs.	61
Welding and Tempering Springs—Mending a Spring with a Broken Ear.	64
How to Mend Wagon Springs.	65
Welding Springs.	66
How to Weld a Buggy Spring.	66
Welding Shaft Irons for Buggies.	67

Welding a Collar on Round Iron.	69
Welding a Round Shaft.	70
Welding.	70
Welding Shafts to an Exact Length.	73
Welding a Heavy Shaft.	74
Welding Boiler Flues.	74
Making a Weld on a Heavy Shaft	75
Welding Angle Iron.	76
Welding Collars on Round Rods.	79
Shall Sand be Used in Welding?	80
Fluxes or Welding Compounds for Iron or Steel.	80
Sand in Welding—Facing Old Hammers.	81
Composition for Welding Cast-Steel.	82
Brazing Cast-Iron.	82
Brazing Ferrules.	83
Brazing a Ferrule.	83
Brazing.	84
Brazing an Iron Tube.	85
Brazing a Broken Crank	85
Brazing with Brass or Copper.	86
Soldering Fluids.	86
CHAPTER IV. STEEL AND ITS USES.	**87**
Tempering, Hardening, Testing.	87
Testing.	92
Hints Regarding Working Steel.	93
The Warping of Steel During the Hardening Process.	94
Tempering Steel.	96
Another Method of Tempering Steel.	97
Working Steel.	99

Working and Tempering Steel.	99
Tempering Steel.	101
Tempering Small Articles.	103
Tempering Steel.	103
Tempering Steel with Low Heat.	104
To Temper Steel Very Hard.	104
Hardening Steel.	105
Case-Hardening Steel or Iron.	105
How Damascus Sword Blades Were Tempered.	106
To Harden Steel.	106
Tempering Steel Springs.	106
Size of Spring.	107
Tempering Small Tools.	110
To Temper Small Pieces of Steel.	111
Hardening Thin Articles.	111
Sword Blades.	111
To Temper Steel on One Edge.	112
Heating to a Cherry-Red—Points in Tempering.	112
Brine for Tempering.	112
A Bath for Hardening Steel.	114
The Lead Bath for Tempering.	114
Hardening Small Tools.	115
Hardening in Oil vs. Hardening in Water.	115
Tempering Plow Points.	116
Tempering Blacksmiths' Tools.	116
Softening Chilled Castings.	116
To Harden Cast-Iron.	116
Brass Wire—How Should it be Tempered for Springs?	117
To Harden Steel Cultivator Shovels.	117

CHAPTER V. FORGING IRON. — 118

- Hand Forgings. — 118
- Making a T-shaped Iron. — 120
- Another Method of Making a T-shaped Iron. — 121
- Forging Stay Ends and Offsets. — 122
- Making an Eye-Bolt. — 126
- Forging a Turn-Buckle. — 127
- Making a Cant-Hook. — 130
- How Forks are Forged. — 131
- Five Methods of Making one Forging. — 134
- Making Offsets. — 139
- To Make a Square Corner. — 140
- Making a Square Corner. — 141
- The Breaking of Step-Legs. — 141
- Making a Thill Iron. — 142
- Making a T-shaped Iron. — 143
- Making a Step-Leg. — 145
- Forging a Head-Block Plate for a Double Perch. — 147
- Forging a Dash Foot. — 148
- How to Make a Slot Circle. — 149
- Forging a Clip Fifth Wheel. — 150
- Method of Making Fronts for Fifth Wheels. — 152
- Making a Fifth-Wheel Hook or a Pole Stop. — 154
- Making a Shift Rail. — 156
- Making Shifting Rail Prop Irons by Hand. — 158
- Getting out a Solid King-Bolt Socket. — 161
- Heading Bolts. — 164
- Heading Bolt Blanks. — 165
- Bending a Cast-Steel Crank Shaft for a Ten Horse-Power Engine. — 167

Making a Clevis.	168
Crank Shafts for Portable Engines.	170
Forging a Locomotive Valve Yoke.	171
Forging a Locomotive Valve Yoke.	173
Forging a Locomotive Valve Yoke.	174
Forging a Locomotive Valve Yoke.	177
Forging a Locomotive Valve Yoke.	178
Forging a Locomotive Valve Yoke.	178
Forging a Locomotive Valve Yoke.	179
Defect in Engine Valve.	187
CHAPTER VI. MAKING CHAIN SWIVELS.	**189**
Making a Log-Chain Swivel.	189
Making a Log-Chain Swivel.	191
Making A Log-Chain Swivel	194
Making a Swivel For a Log Chain.	196
Making a Swivel for a Log Chain.	198
PLAN 5.	198
Making a Swivel For a Log Chain.	200
Making a Swivel For a Log Chain.	202
CHAPTER VII. PLOW WORK.	**203**
Points in Plow Work. No. 1.	203
Points In Plow Work. No. 2.	206
Some More Points About Plows.	206
Making a Plow Lay.	207
Laying A Plow.	208
Polishing Plow Lays and Cultivator Shovels.	211
Laying a Plow. Plan 1.	211
Laying a Plow. Plan 2.	213
Laying a Plow. Plan 3.	213

Laying, Hardening and Tempering Plows.	216
Laying Plows.	218
Making a Plowshare.	220
How to Sharpen a Slip-Shear Plow Lay.	222
Welding Plow Points.	223
How to Put New Steel Points on Old Plows.	224
Pointing Plows.	226
Tempering Plow Lays and Cultivator Shovels.	228
Sharpening Listers.	228
Notes on Harrows.	229
Making a Bolt-Holder and a Plowshare.	232
Making a Grubbing Hoe.	233
Making a Grubbing Hoe.	236
Forging a Garden Rake.	237
Making a Double Shovel Plow.	237
Pointing Cultivator Shovels.	242
Pointing Cultivator Shovels.	243
Pointing Cultivator Shovels.	244

PREFACE.

Vol. I. of this series was devoted to a consideration of the early history of blacksmithing, together with shop plans and improved methods of constructing chimneys and forges.

Vol. II. was, for the most part, given up to a consideration of tools, a great variety of which were described and illustrated.

In Vol. III. the subject of tools is continued in the first and second chapters, after which the volume is devoted chiefly to a description of a great variety of jobs of work. This volume is, therefore, in many respects, the most valuable thus far of the series, as it shows how the improved tools described in Vols. I. and II. can be used practically.

Vol. IV. will continue the topic of jobs of work and complete the series.

CHAPTER I.

BLACKSMITH TOOLS.

THEIR PRESERVATION.

In continuing the construction of blacksmith's tools from Vol. II. some general directions for their care and preservation will not be out of order, as even the best tools soon become useless if they are not well cared for. The following valuable hints on their preservation will be appreciated by every mechanic who has a desire to make his tools last as long as possible, and who wishes to have them always in good condition:

Wooden Parts.—The wooden parts of tools, such as stocks of planes and the handles of chisels, are often made to have a nice appearance by French polishing, but this adds nothing to their durability. A much better plan is to let them soak in linseed oil for a week and rub them with a cloth for a few minutes every day for a week or two. This produces a beautiful surface and exerts a solidifying and preservative action on the wood.

Iron Parts.—*Rust Preventatives*—1. Caoutchouc oil is said to have proved efficient in preventing rust, and it has been used by the German army. It only requires to be spread with a piece of flannel in a very thin layer over the metallic surface and allowed to dry up. Such a coating will afford security against all atmospheric influences and will not show any cracks under the microscope after a year's standing. To remove it, the article has simply to be treated with caoutchouc oil again and washed after twelve to twenty-four hours.

2. A solution of India rubber in benzine has been used for years as a coating for steel, iron and lead, and has been found a simple means of keeping them from oxidizing. It can be easily applied with a brush and as easily rubbed off. It should be made about the consistency of cream.

3. All steel articles can be perfectly preserved from rust by putting a lump of freshly-burnt lime in the drawer or case in which they are kept. If the things are to be moved, as a gun in its case, for instance, put the lime in a muslin bag.

This is especially valuable for specimens of iron when fractured, for in a moderately dry place the lime will not need renewing for many years, as it is capable of absorbing a large amount of moisture. Articles in use should be placed in a box nearly filled with thoroughly slaked lime. Before using them rub well with a woolen rag.

4. The following mixture forms an excellent brown coating for preserving iron and steel from rust: Dissolve two parts crystallized iron of chloride, two of antimony of chloride and one of tannin in four of water, and apply with sponge or rag and let dry. Then another coat of paint is applied, and again another, if necessary, until the color becomes as dark as desired. When dry it is washed with water, allowed to dry again and the surface polished with boiled linseed oil. The antimony chloride must be as nearly neutral as possible.

5. To keep tools from rusting, take one-half ounce camphor, dissolve in one pound melted lard; take off the scum and mix in as much fine black lead (graphite) as will give it an iron color. Clean the tools and smear with this mixture. After twenty-four hours rub clean with a soft linen cloth. The tools will keep clean for months under ordinary circumstances.

6. Put one quart freshly slaked lime, one-half pound washing soda and one-half pound soft soap in a bucket, and sufficient water to cover the articles; put in the tools as soon as possible after use, and wipe them next morning, or let them remain until wanted.

7. Soft soap, with half its weight in pearlash, one ounce of mixture in one gallon of boiling water, is in everyday use in most engineers' shops in the drip-cans used for turning long articles bright in wrought-iron and steel. The work, though constantly moist, does not rust, and bright nuts are immersed in it for days, till wanted, and retain their polish.

8. Melt slowly together six or eight ounces of lard to one ounce of resin, stirring until cool; when it is semi-fluid it is ready for use. If too thick it may be further let down by coal oil or benzine. Rubbed on bright surfaces ever so thinly, it preserves the polish effectually and may readily be rubbed off.

9. To protect metal from oxidation, polished iron or steel, for instance, it is requisite to exclude air and moisture from the actual metallic surface; therefore, polished tools are usually kept in wrappings of oil-cloth and brown paper, and thus protected they will preserve a spotless face for an unlimited

time. When these metals come to be of necessity exposed, in being converted to use, it is necessary to protect them by means of some permanent dressing, and boiled linseed oil, which forms a lasting covering as it dries on, is one of the best preservatives, if not the best. But in order to give it body, it should be thickened by the addition of some pigment, and the very best, because the most congenial of pigments, is the ground oxide of the same metal, or, in plain words, rusted iron reduced to an impalpable powder, for the dressing of iron and steel, which thus forms the pigment of oxide paint.

10. Slake a piece of quicklime with just enough water to crumble in a covered pot, and while hot add tallow to it, and work into a paste, and use this to cover over bright work; it can be easily wiped off.

11. Olmstead's varnish is made by melting two ounces of resin in one pound of fresh, sweet lard, melting the resin first and then adding the lard and mixing thoroughly. This is applied to the metal, which should be warm, if possible, and perfectly clean; it is afterward rubbed off. This has been well proved and tested for many years and is particularly well suited for Danish and Russian oil surfaces, which a slight rust is apt to injure very seriously.

Rust Removers.—1. Cover the metal with sweet oil, well rubbed in, and allow to stand for forty-eight hours; smear with oil applied freely with a feather or with a piece of cotton wool after rubbing the steel. Then rub with unslaked lime reduced to as fine a powder as possible.

2. Immerse the article to be cleaned for a few minutes, until all the dirt and rust are taken off, in a strong solution of potassium cyanide, say about one-half ounce in a wineglass of water; take it out and clean it with a toothbrush with a paste composed of potassium cyanide, castile soap, whiting and water mixed into a paste of about the consistency of thick cream.

BENCH TOOLS.

The tool shown in Fig. 1 is very convenient where there is much bundle iron to open, as it is made heavy enough so that any ordinary band can be easily cut with it at one blow. It has an eye large enough to admit a small sledge handle, and the handle should be made of good hickory with some surplus stock near the eye, as it is liable to get many bruises from careless handling and mis-blows.

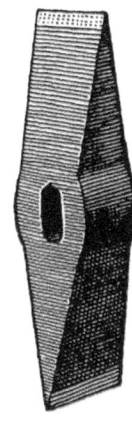 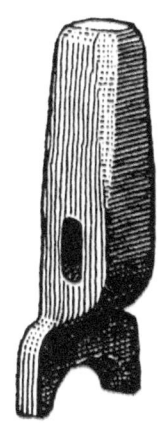

FIG. 1—HATCHET FOR OPENING BUNDLE IRON. MADE OF 1 5/8 SQUARE STEEL.

FIG. 2—COLLARING TOOL FOR SHOULDERING DOWN ROUND IRON. MADE OF 1¼ SQUARE STEEL.

It should have a little less blunt edge than a cold chisel and be tempered a "pigeon-blue," if it is made of good steel; but if it is made of the fancy brands the temper must be a matter of experiment.

For shouldering down round iron or steel to form a collar or neck, there is no tool that is any better than that shown in Fig. 2. The concave should be of a size to fit the circumference of the bar to be worked or larger.

The cut does not show the cutting part quite plainly; the edge all the way around the hollow should be flat on the inside and rounded out on the other side the same in section as Fig. 8.

Fig. 3 is simply a good handy size for a light flatter. It is about 5½ inches high. There is a great advantage in having a flatter light, not only because it is easier handled, but because it is more efficient. When a flatter is too heavy in proportion to the weight of the sledge it absorbs more force than it gives down. It kicks.

It spends its elasticity in reacting against the sledge, instead of letting the blow through it and delivering it to the work on the other side. It is all nonsense to suppose that big flatters are best on big work. It is not the work that governs the size of hand tools, it is the power of the men who are to deliver the blows.

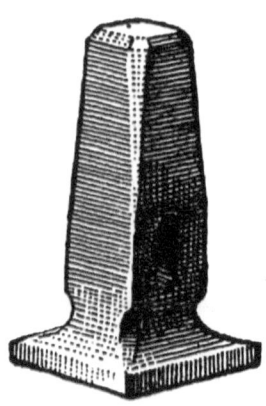

FIG. 3—LIGHT FLATTER FOR FINISHING FLAT IRON. MADE OF 1 ¾ SQUARE STEEL. FACE 2 ½ INCHES SQUARE.

Fig. 4 is a tool that does not feel as good in the hand, and is not quite as nice to handle. It some way does not hang as well as a flatter, but it is a tool that should be used in the formation of all inside corners, for it is a deadly enemy to cold shuts and broken fibers, which are the vital seeds of death in any work of iron in which they find lodgment.

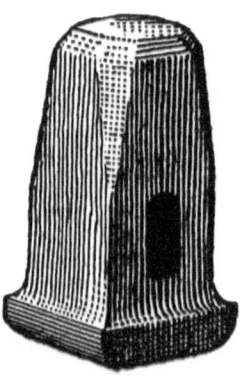 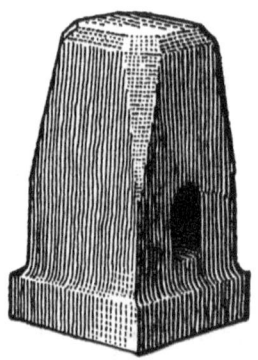

FIG. 4—ROUND-CORNERED SET HAMMER. MADE OF 1¾ SQUARE STEEL.

FIG. 5—HEAVY FLATTER FOR STRAIGHTENING COLD IRON. MADE OF 2-INCH SQUARE STEEL.

The heavy flatter, Fig. 5, for straightening cold iron, is made very strong, and a sledge must be used with it proportionate to its weight.

There is not such particular need of activity, spring, and haste in getting a blow on cold iron as there is on hot, and blows that count in bending or straightening, are slow and solid. Steel rails are straightened under a press.

If this flatter is not made very strong it will soon crystallize and break in the weakest place across the eye.

The tools, Figs. 6, 7, 8, are for the purpose of siding-down work or making offsets, leaving good shoulders standing up, without having to use the backing hammer. There is a tendency to make tools heavier than is necessary simply to perform the office in blacksmithing that the jointer plane does in carpentering. The carpenter jacks off the rough stock and then smooths up with his "jointer." In dressing tools a heavy large-faced hammer is used by some first-class tool dressers. I have known them to call it the "jineter."

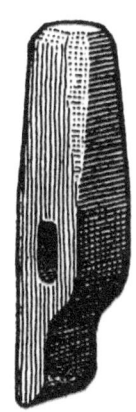
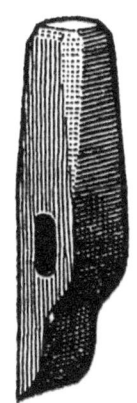

FIG. 6—LARGE SIDING-DOWN TOOL. MADE OF 1½ SQUARE STEEL.

FIG. 7—SMALL SIDING-DOWN FULLER. MADE OF 1¼ SQUARE STEEL.

FIG. 8—SMALL SIDING-DOWN CHISEL. MADE OF 1¼ SQUARE STEEL.

The siding-down tool, Fig. 8, need not be wider than a man can sink an eighth of an inch into hot iron or steel at a blow. When the impression is deep enough, or if, in crossing wide iron, it gets crooked sidewise, the wider bitted one shown in Fig. 6 can be used to make the impression straight and uniform, and afterwards the siding-down fuller, Fig. 7, may be used. On a large amount of work these tools suffice; but where there is much wide iron to work it will pay to have a wide fuller, the width, say two and one-half inches, of Fig. 6.

BLACKSMITH TONGS.

The blacksmith who will do his work well and quickly, whether on carriage work or the ordinary work of the country shop, must be well supplied with tongs, and they well made. It is no uncommon thing to see a man working at the forge depending upon two or three tongs for holding all kinds of work. If the jaw opens too wide it is heated and a blow from the hammer closes it; if too narrow the same operation is gone through to open it; this makeshift business costs dear, and brands the workman as a botch.

A complete list of tongs for one man might not be a complete list for another, as some workmen are particular as regards specialties, but an assortment that comprises those that should be on every bench consists of two pairs of tongs for ⅛-inch iron, two pairs for ¼-inch iron, two pairs for ⅜-inch iron, two pairs for ½-inch iron, and one pair for each succeeding one-eighth of an inch up to 1 ¼ inches and above that a pair for each succeeding quarter inch up to the limit of size.

Blacksmiths, as a rule, prefer to make their own tongs. For these they should use Lowmoor or Burden's "best." Drill all holes, instead of punching, and be careful to see that the face of the jaws are parallel when closed to the required size; jagging or otherwise roughing the face of the jaw is an unnecessary operation, for if the tongs work easily and true, as they should, they will hold the iron without extra pressure. If the jaws wobble or twist the fault is at the joint and should be corrected. The blacksmith who stands all day at the forge working with poor tongs will find, when night comes, the hand that held the tongs is much more wearied than the one that held the hammer.

HOW TO MAKE A PAIR OF COMMON TONGS.

I will describe my method of making a pair of common tongs, which is so simple that any blacksmith can follow it. I take a piece of ½ x 1 ¼ inch iron,

14 inches long, and draw down the ends as shown in Fig. 9. Fig. 10 shows a side view. Then split as shown in Fig. 11 and draw the handles to the proper shape. Punch the holes, rivet, and the tongs are completed and can be shaped to suit your own notion.—*By* J. M. W.

HOW TO MAKE A PAIR OF COMMON TONGS. FIG. 9—SHAPE TO WHICH "J. M. W." WOULD DRAW THE IRON.

FIG. 10—SIDE VIEW OF FIG. 9.

FIG. 11—HOW THE IRON IS SPLIT.

TOOLS FOR FARRIER WORK.

Fig. 12 shows a shoe-spreader. To make it take ¾ or ⅞-inch square Norway iron, shoulder and turn down as shown at A and B. Fig. 13 shows a side view of B of Fig. 12.

For *C* use three-fourths rod with thread up to the jaw *A* and riveted through it. File notches in points so that they will not slip.

This tool is very useful and can be used to spread a shoe that has been on two or three weeks, or when only one side is nailed. D, of Fig. 12, is marked in inches, so that one can tell the exact distance the shoe has been spread.

Fig. 14 shows a farrier's pick for removing dirt and gravel. It is made of ½-inch steel and has a hole in the end that it may be hung on box.

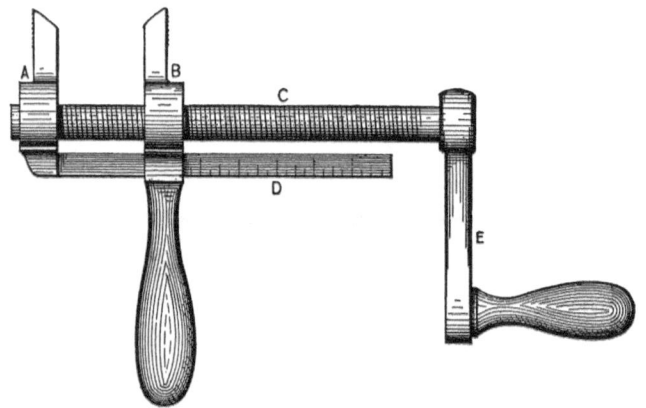

FIG. 12—SHOE SPREADER COMPLETE.

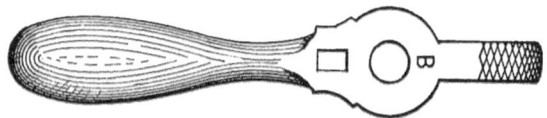

FIG. 13—SIDE VIEW OF B, FIG. 12.

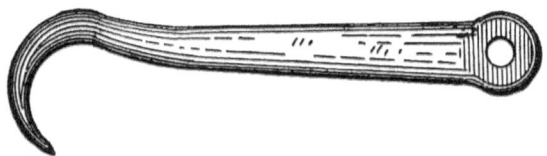

FIG. 14—FARRIER'S PICK.

Fig. 15 is a farrier's corn-cutting tool.

It is made of ¼-inch round steel and has the point ground to a sharp diamond tip. It is worth its weight in gold to any horseshoer. The handle is that of a farrier's broken knife.

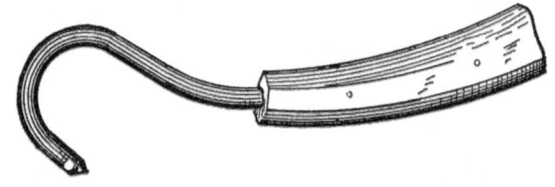

FIG. 15—FARRIER'S CORN-CUTTING TOOL.

As every blacksmith is acquainted with the clinch block it does not need illustrating.

I have mine made rounded to fit the shoe and with a groove to fit outside of the crease in the shoe, and runs up the side of the shoe, the idea being to hold nails that are sunk too deep for corn block.

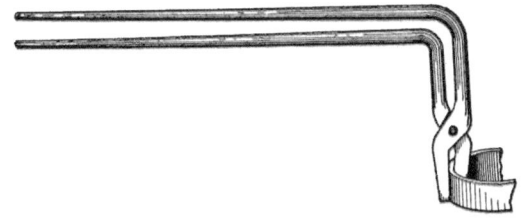

FIG. 16—HANDY TONGS FOR HANDLING WAGON TIRES.

Fig. 16 shows a pair of fire-tongs made like the ordinary fire-tongs but having the handles bent four or five inches above the jaws. These tongs are to be used in cases where it is desirable to keep the hands away from the fire. Especially are they handy to use in handling wagon tires.

Fig. 17 shows tool used in sharpening toe calks.

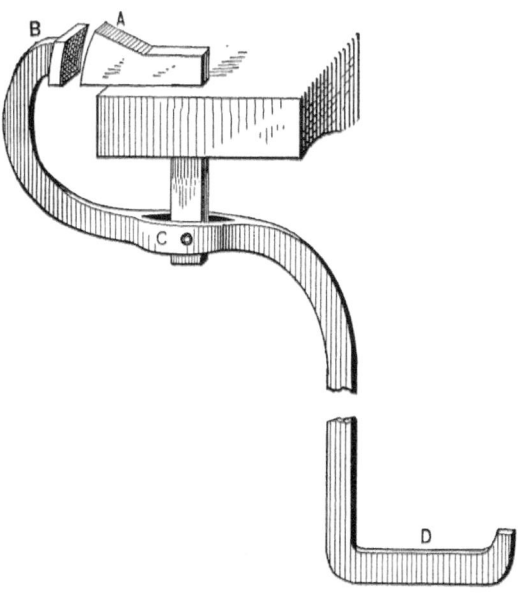

FIG. 17—TOOL USED IN SHARPENING TOE CALKS.

The part *A* is made of tool steel, and is swaged same as an ordinary bottom swage, raised at *A*, and slightly rounded, so that the toe of the shoe will stand out. The part *B, C, D* is made of one by one-fourth inch iron, and to the shape shown in Fig. 17. It should be long enough so that the smith can keep his heel on the floor and place his toe on *D*. No weights are necessary to raise the jaw *B*.

This tool has the merit of simple construction. Of course it is intended for use on end of the anvil. Fig. 18 shows a pair of tongs that come very handy for holding horseshoes.

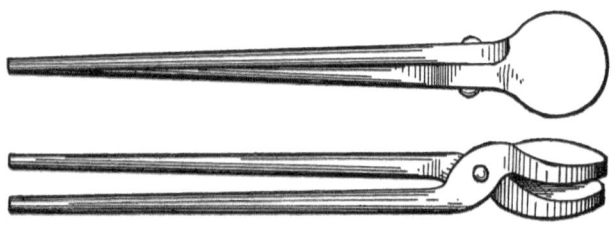

FIG. 18—TONGS FOR HOLDING HORSESHOES.

These are so well known as to need no explanation. Fig. 19 shows a handy yoke puller. It is made of ⅜-inch round iron and hinged and riveted. One point is turned up to fit in hole of yoke, the other is rounded to fit clip, as seen in cut.

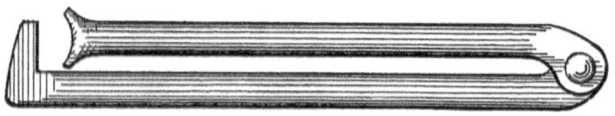

FIG. 19—YOKE PULLER.

The plow clamp shown in Fig. 20 is made of 3-4 x 7-16 horseshoe bar, turned and welded. It fits the share edgewise. The space is for 5-16 bolt, and bolts to share with cam.

FIG. 20—PLOW CLAMP.

The bolts keep the share from springing when being sharpened. Fig. 21 shows the tongs I use for holding cultivator shovels.

FIG. 21—TONGS FOR HOLDING CULTIVATOR SHOVELS.

The under piece has forks that pass on either side of the casting on the shovel or ball tongue. The upper jaw is similar to that of ordinary tongs, except that it is a little shorter than the forks. The handles arc bent a little so that the ball tong point stands nearly straight.

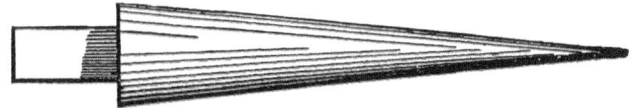

FIG. 22—HORN FOR WELDING FERRULES.

Fig. 22 shows a horn for welding ferrules and small bands, also for rounding the same. It is made to fit anvil and is one and one-half inches at bottom and tapers to a point. The length of horn is eight inches.

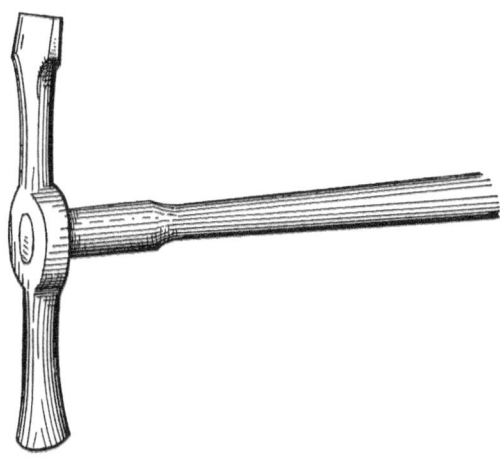

FIG. 23—LIGHT RIVETING HAMMER.

Fig. 23 shows a light riveting hammer. It is made of five-eighths steel. Draw and make like ordinary hammer, except that the handle should be very light and elastic.

This hammer is very handy in riveting light castings, light welds, etc.

A plow hammer is seen in Fig. 24. It weighs two pounds, and has the pene set lengthwise with the handle and enables the smith to weld in throat of plow.

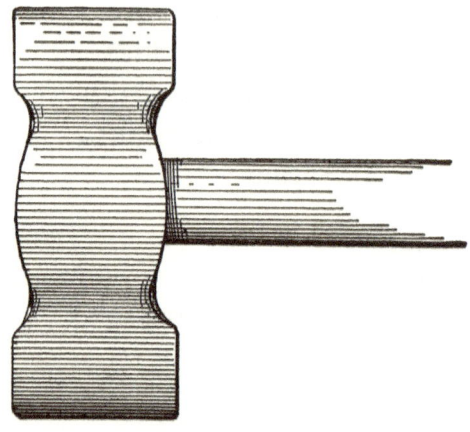

FIG. 24—LIGHT PLOW HAMMER.

Fig. 25 shows a turning hammer with two faces; one is made rounding for concaving shoes, it is also handy for drawing any kind of iron. It weighs from two and one-half to three pounds.

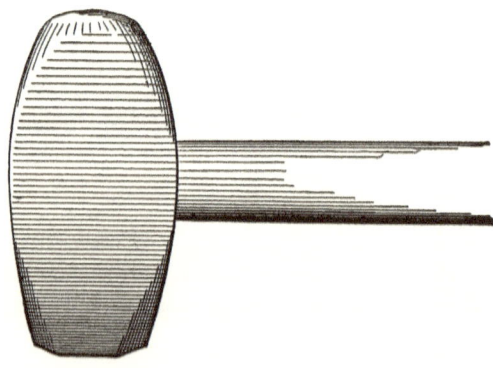

FIG. 25—TURNING HAMMER.

Fig. 26 shows a round chisel. It is made similar to the ordinary handle chisel, except that it is made round in two sizes—1 and 1 ¾ inches—for cutting holes in wagon plates, roller plates, etc.

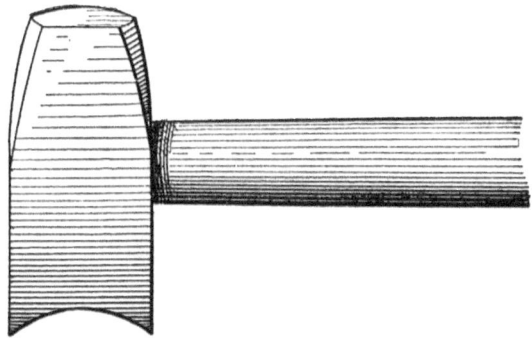

FIG. 26—ROUND CHISEL.

A hoop set-hammer is seen in Fig. 27. It is made lighter than the ordinary set-hammer and tapering on sides only. It is used for band hoops or any kind of band drawing.

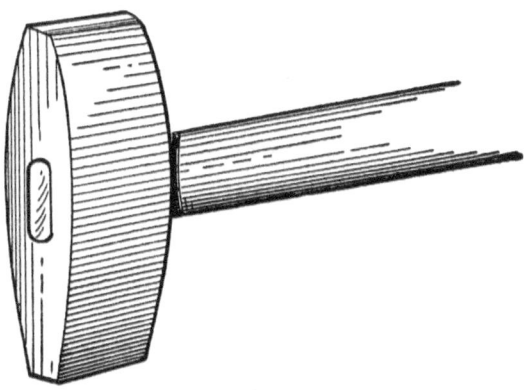

FIG. 27—HOOP SET-HAMMER.

Fig. 28 shows a singletree clip wedge. It is made of ½ x 2-inch iron with a groove on bevel side, and is used to draw tight single and doubletree clips. In putting them on, slip on clip, drive in wedge tight down, and cool, drive out the wedge and the job is done.

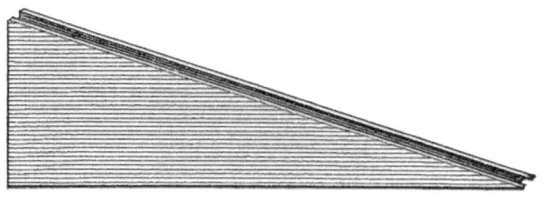

FIG. 28—SINGLETREE CLIP WEDGE.

Fig. 29 shows a vise tool for holding short bolts. It is made of 2 x 2 ½-inch iron bent square, and has three grooves cut for three sizes of bolts, shouldered off, and riveted at bottom. This tool is handy in cutting threads on plow, or any short bolts, or in working nuts on same.

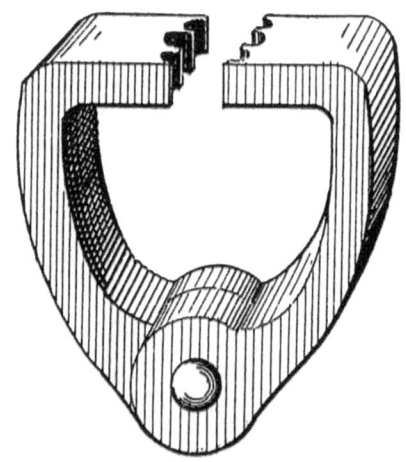

FIG. 29—VISE TOOL FOR HOLDING SHORT BOLTS.

Fig. 30 shows a heavy wrench or bending tool. This tool is so common among our smiths as to need no description.

FIG. 30—HEAVY WRENCH OR BENDING TOOL.

Fig. 31 shows a forge crane. The upright post is made of 1 ½ x ½-inch iron and shouldered at *C;* draw around to five-eighths or three-fourths; place collar on at shoulder. Bore a hole close to the forge to receive *C.*

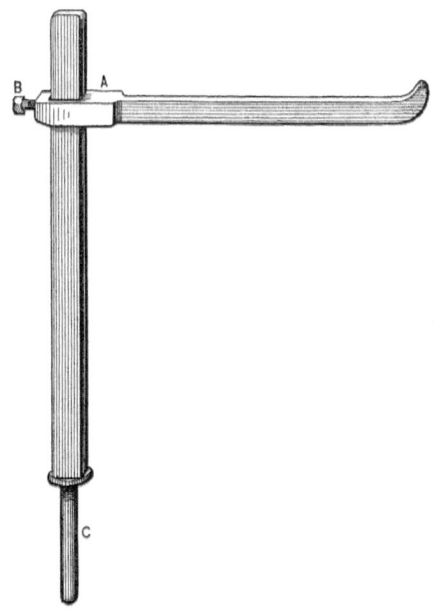

FIG. 31—FORGE CRANE.

The crane is made of inch square iron and should slide easily, and is held in position by the set-screw at B. The upright should extend seven or eight inches above the forge. If there is no floor in the shop then drive or set a post level with the dirt. Blacksmiths will find this to be a great labor-saving tool.

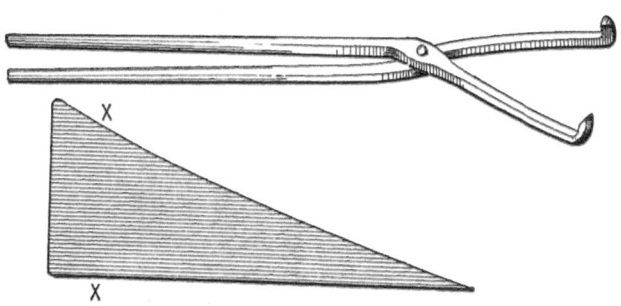

FIG. 32—TONGS FOR PLOWSHARE.

Fig. 32 shows a pair of handy tongs for plowshares. The jaws are made to fit top and bottom of share, being turned to fit the bevel as shown at *x x* of share. These tongs are used in either welding or sharpening.

Fig. 33 shows a clip for making round clips. It is made to fit the anvil and can be made in any size.

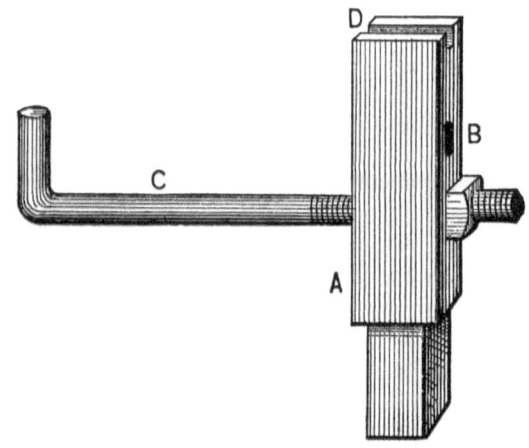

FIG. 33—TOOL FOR MAKING ROUND CLIPS.

C is the gauge, *B* the hole, and *D* the groove in top. Cut the iron to the right length for clip wanted, cut threads on both ends, heat and run through the hole *B*, gauge by *C*, and bring the end over to *D*, tap down gently until true.

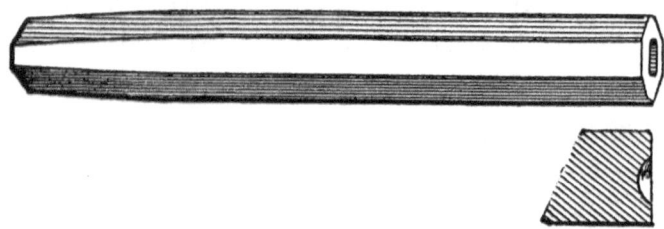

FIG. 34—TOOL FOR ROUNDING RIVET HEADS.

Fig. 34 shows a tool for rounding rivet heads. It is made of one-half inch steel shaped like a punch. Make a tool to the shape that you want the head of the rivet to be when finished. Heat the steel and place it in the vise, then drive the special tool, or rivet head, into the steel until it is sunk enough. Then dress up and temper to a light blue.—*By* Rex.

A TOOL FOR HOLDING PLOW BOLTS.

I will try to give a description of a handy tool for holding plow bolts.

The piece *A* shown in Fig. 35 is made of ⅜ or ½-inch iron, and is about 20 inches long.

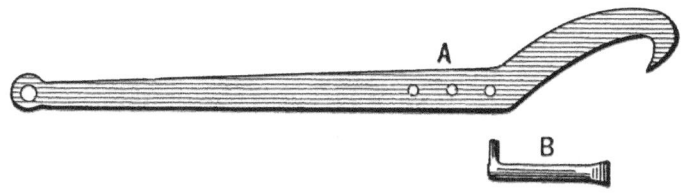

A TOOL FOR HOLDING PLOW BOLTS. FIG. 35—SHOWING HOW THE TWO PIECES ARE MADE.

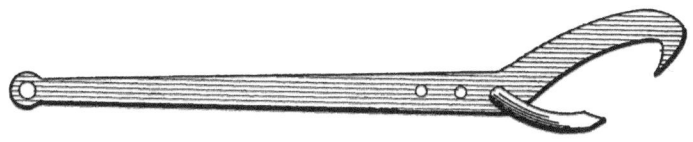

FIG. 36—SHOWING THE TOOL COMPLETED.

It has holes in it into which the piece *B* can be inserted and moved forward or backward so as to catch any bolt. The piece *B* is made of steel with a rounded end to fit in the hole in *A*. The other end is made like a cold chisel in order to catch the bolt. Fig. 36 represents the tool ready for use.—*By* A. G. Bunson.

TONGS FOR HOLDING PLOW POINTS.

I have a pair of tongs for holding plow points while sharpening or laying, that are simple, easily made, and I like them far better than any other tongs for the purpose that I ever saw. I forged them from a one and a quarter inch square bar just like ordinary straight jawed tongs. The edges are about two inches long (not longer), quite heavy, with one-half inch handles. After they were finished I heated them and then caught them edgewise in the vise and bent them, just at the rivet, to an angle of nearly forty-five degrees, and I find they never slip or work off, but answer every purpose.—*By* Earles J. Turner.

A TOOL FOR HOLDING PLOW BOLTS.

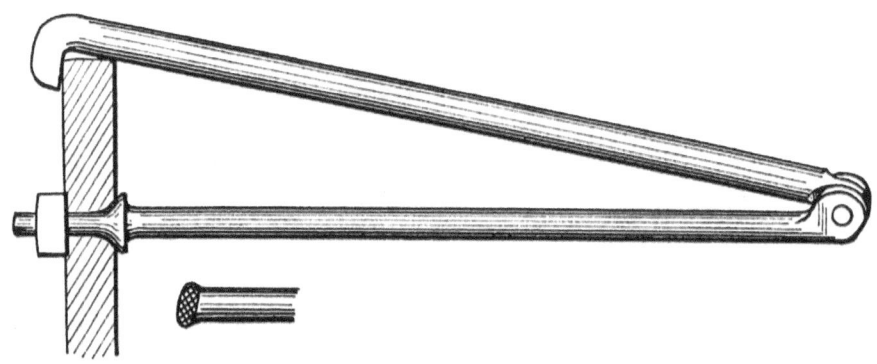

FIG. 37—TOOL MADE BY "H. H. K." FOR HOLDING PLOW BOLTS.

A handy little tool which I use to prevent plow bolts from turning when the wrench is on the nut is shown in Fig. 37. The tool is one that will be appreciated by every smith who does plow work. It is made of 5/8-inch round iron, but having steel at one end which is cross-cut as shown in the illustration.—*By* H. H. K.

A TOOL FOR HOLDING SLIP-SHEAR PLOWS IN SHARPENING.

To make a tool for holding slip-shear plows in sharpening them, take 5/8-inch round iron, cut off two pieces, making each 2 ½ feet long, and bend one in the middle and weld the ends as shown in Fig. 38, so that a plow bolt will fit in and slip along.

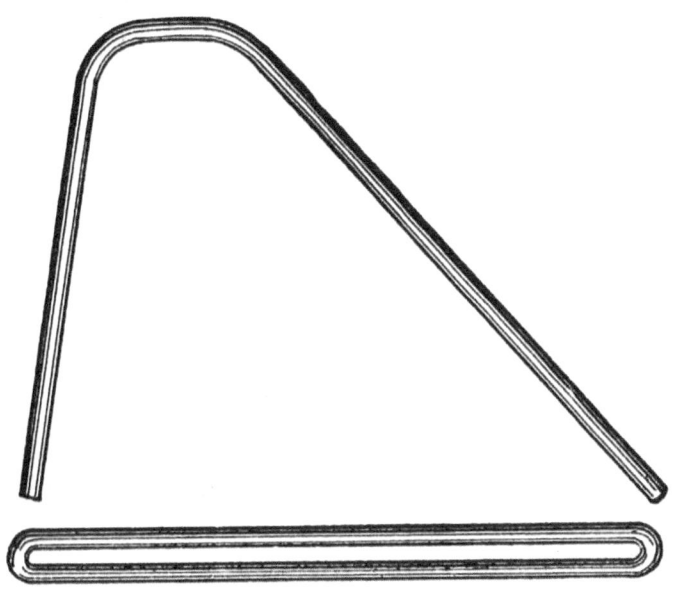

A TOOL FOR HOLDING SLIP-SHEAR PLOWS IN SHARPENING. FIG. 38—SHOWING HOW THE TWO PIECES ARE SHAPED.

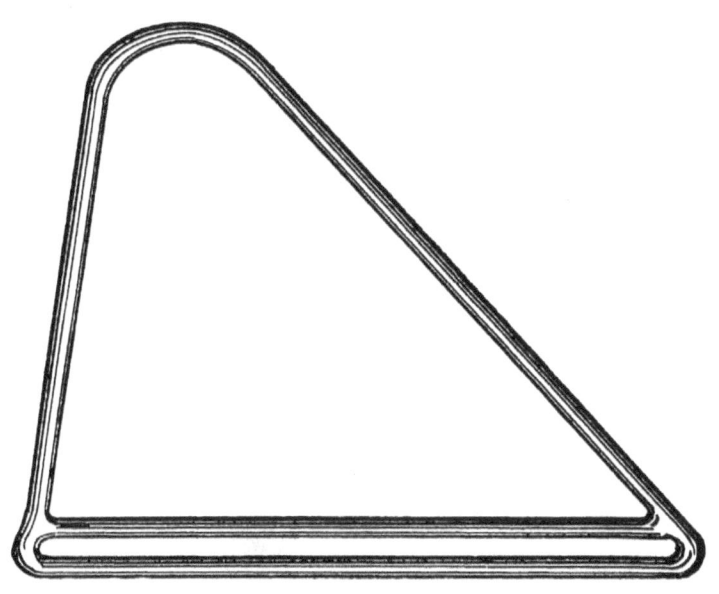

FIG. 39—SHOWING THE TOOL AS FINISHED.

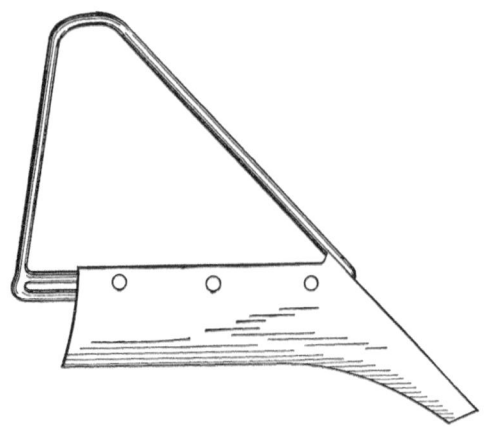

FIG. 40 SHOWING THE TOOL FASTENED TO A PLOWSHARE.

I then bend the other piece as shown in the cut, then weld the ends of this piece to those of the other, and the tool is finished as shown in Fig. 39. In Fig. 40 the tool is shown fastened to a plowshare with plow bolts. This tool will hold either right or left-hand plows.—*By* A. G. B.

MAKING A PLOW BOLT CLAMP.

To make a plow bolt clamp take a piece of steel 14 inches long, and 5/8-inch square; make a two-pronged claw to fit the bolt-head on one end and draw the other to go into the wooden handle marked *A*, Fig. 41. Then draw a piece of 1 ¾ x ½ inch iron to an edge and bend two inches to a right angle at *B*. Punch a square hole to fit the steel, two inches from the bend.—*By* G. W. P.

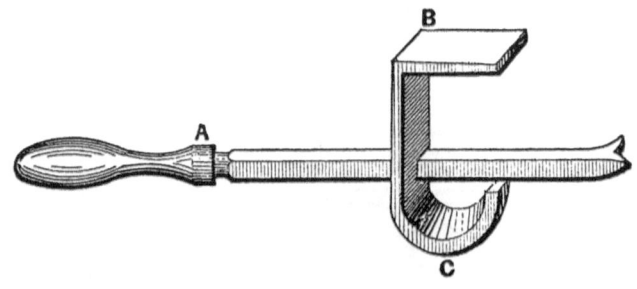

FIG. 41—A PLOW-BOLT CLAMP AS MADE BY G. "W. P".

TONGS FOR HOLDING PLOW BOLTS.

This is a very handy tool and one which no shop should be without. With this tool a bolt in the lay can be held with one hand while the other is free to remove the burr. I consider it the only successful tool ever invented for this purpose.

The jaw *A*, Fig. 42, is five inches in length, while *B* is four and one-half inches long. The point *C* is made of steel and welded to *B*, and must be tempered hard. It is made with a sharp point like a chisel or screw point. The handles are two feet long and of five-eighths inch iron. The jaws are of three-fourths inch square iron.

To remove a bolt from a plow-lay with this tool place the point *C* on the bolt head, and let the jaw *A* come in any convenient place on the other side of the lay, grip tightly and the bolt will be held tight while the nut is being removed.

FIG. 42—IRON TONGS FOR HOLDING BOLT HEADS, AS MADE BY E. K. WEHRY.

If the nut be rusted on and hard to turn, then with a sharp chisel cut across the bolt head same as a screw head. Then place C in the cut and the bolt cannot turn. You will seldom have to do this.—*By* E. K. Wehry.

HOW TO POINT A PLOW.

My plan for pointing a plow is as follows: Make a pair of blacksmith's tongs, somewhat heavier than ordinary tongs, let one jaw be two inches and the other one five inches long.

Make the long jaw very heavy and shaped as shown in Fig. 43; then take a piece of suitable steel and cut out a point the desired shape, and, after shaping and filing the edges, place it on the plow lay and clamp it with the tongs, as shown in Fig. 44; then take a light heat on the point and bend it under, as shown by the dotted lines. Make the point of such a length that when bent under it will lap on the original point from one to two inches.

FIG. 43—POINTING A PLOW, AS DONE BY "H. L. C." THE TONGS.

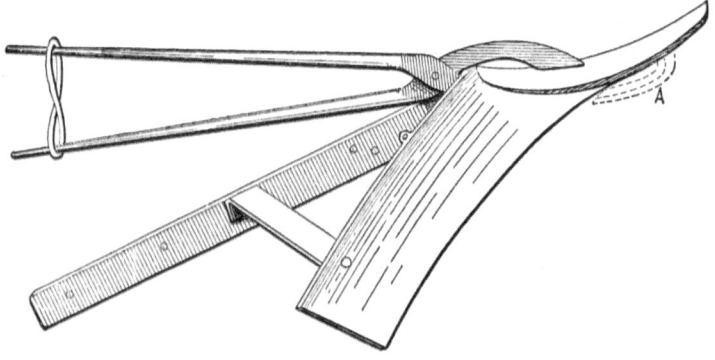

FIG. 44—SHOWING THE CLAMPING AND BENDING PROCESSES.

Then take a thin piece of soft iron and place it between the lap at *A* in Fig. 44—this is to make the point heavier and to cause it to weld better—then take a welding heat on the point, after which the tongs may be taken off and the job finished up. This plan is a great advantage over the old way of drilling the point and share, and riveting the point to hold it in place while taking the first heat. It not only saves time and labor, but it makes a stronger and neater job.— *By* H. L. C.

HINTS FOR PLOW WORK.

Some of the plow manufacturers send out lays that are so badly welded that after being sharpened once or twice they fall away from the landside, and then the farmer blames the blacksmith.

Fig. 45 represents the tongs I use in sharpening lays when there is danger that they will be loosened. Fig. 46 shows how the tongs are used on a slip lay so that the lay and landside will be held together until the welding has been done up to the tongs.

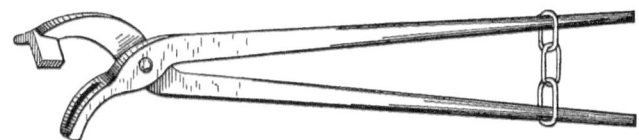

HINTS FOR PLOW WORK. FIG. 45—SHOWING TOOL USED IN SHARPENING LAYS.

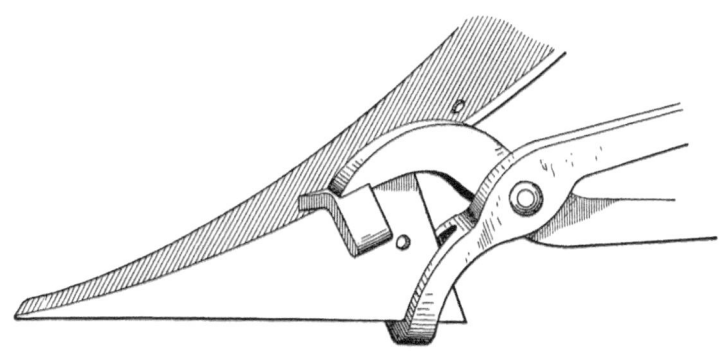

FIG. 46—SHOWING HOW THE TONGS ARE USED.

For the benefit of smiths who have to handle such plows I give a few hints which may prove valuable.

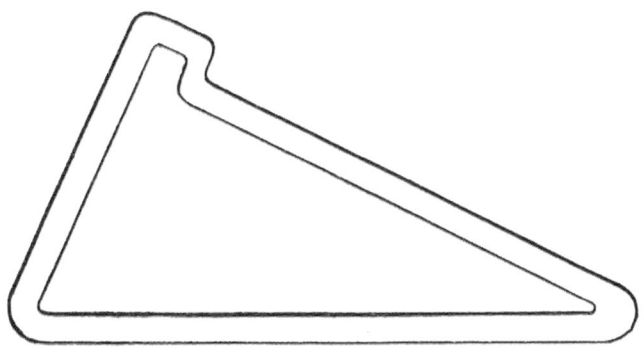

FIG. 47—THE TRIANGLE USED IN WELDING.

Fig. 48 represents a whole landside lay or bar lay. In welding these I use a triangle shown in Fig. 47 and a wedge shown in Fig. 49.—*By* G. W. Predmore.

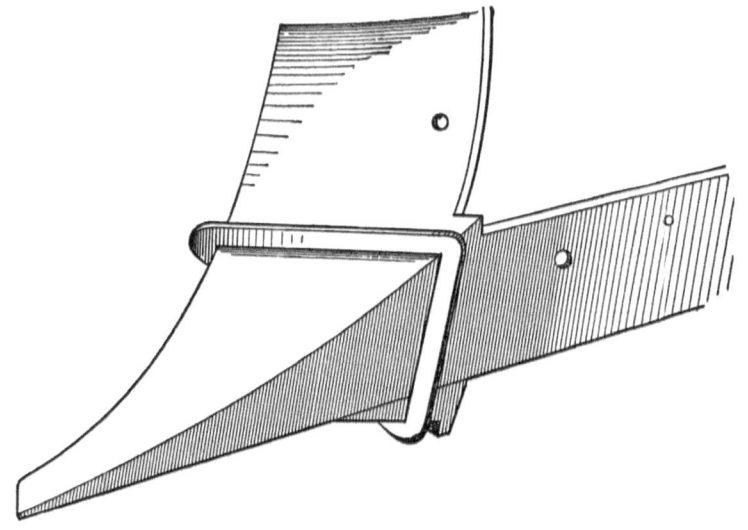

FIG. 48—A LANDSIDE OR BAR LAY.

FIG. 49—THE WEDGE USED IN WELDING.

A TOOL FOR HOLDING PLOWSHARES.

A device invented by me for holding plowshares, which I think is one of the best tools in use for holding plowshares when sharpening or pointing them, is shown by Fig. 50.

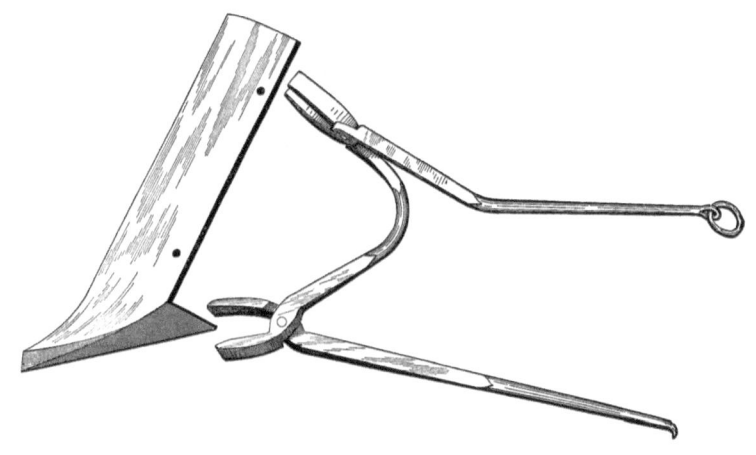

FIG. 50—TOOL MADE BY "G. B." FOR HOLDING PLOWSHARES.

It consists of two pairs of tongs welded together, one holding the bar and the other holding the wing. The tongs holding the wing should have round jaws. When taking hold of the share the handles come together within three inches or so, and the ring on one of the handles is then slipped over the other.—*By* G. B.

CHAPTER II.

WRENCHES.

FORGING WRENCHES.

My way of making a wrench is as follows: For a 3-inch wrench take iron 1x2 for piece *A*, in Fig. 51(Fig. 52 shows it more plainly), and punch a hole at *A* to receive *B*, and take a weld, using the fuller at *C D*.

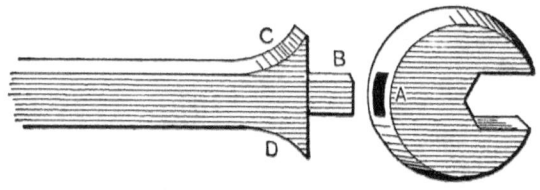

FIG. 51—FORGING A WRENCH.

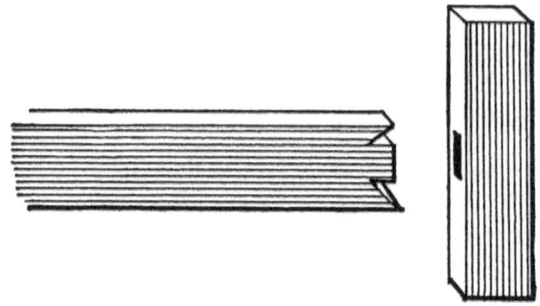

FIG. 52—HOW TO FORM THE ENDS AS SEEN IN FIG. 51.

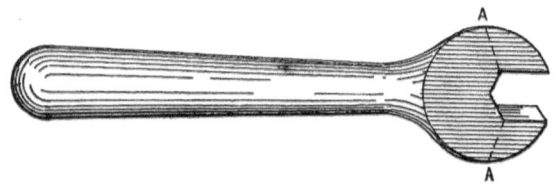

FIG. 53—FINISHED WRENCH.

Fig. 53 shows the wrench complete. It should be very strong at dotted lines *A A*, where the greatest strain comes. Fig. 52 shows how to cut the end to form *C D*, in Fig. 51. Fig. 54 shows how to make small wrenches.

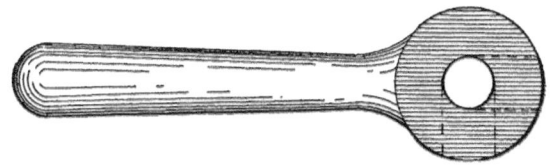

FIG. 54—METHOD OF MAKING SMALL WRENCHES.

Punch hole in the center and cut to any desired angle; see dotted lines.—*By* Southern Blacksmith.

MAKING WRENChES.

My plan of forging a wrench is as follows: Take any piece of iron corresponding in size to the wrench it is desired to make. Bend it as shown in Fig. 55.

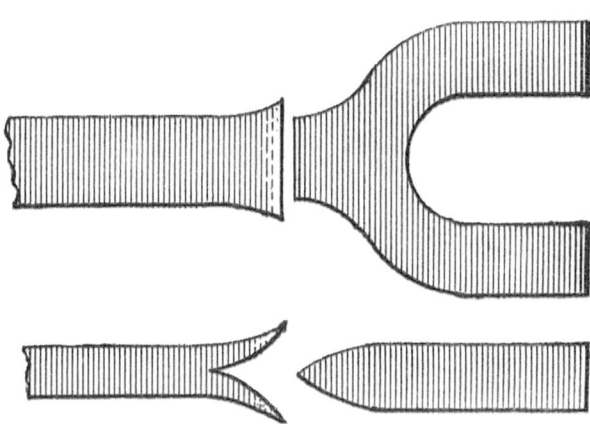

FIG. 55—"J. T. B.'S" METHOD OF MAKING WRENCHES.

After welding, proceed to form into shape. Any practical man, it seems to me, will admit the advantages of the plan shown in the sketch.—*By* J. T. B.

CURVE FOR AN S WRENCH.

I inclose sketches of my way of curving an S wrench. I always make the curve so that the jaws run parallel with each other, as is shown in Figs. 56 and 57.

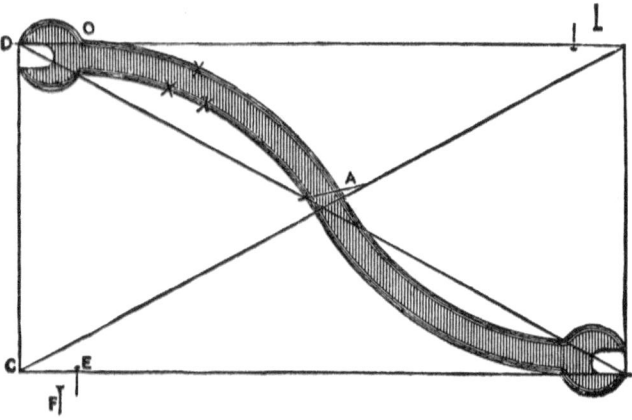

FIG. 56—A CURVED WRENCH.

Fig. 56 is a curved wrench, while Fig. 57 is a straight wrench.

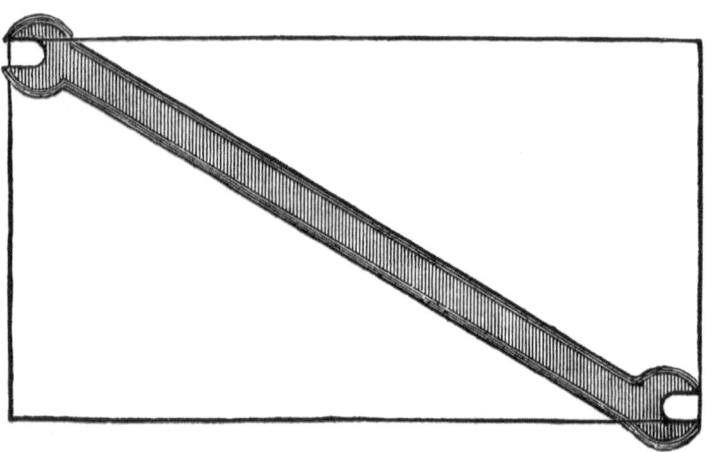

FIG. 57—A STRAIGHT WRENCH.

Both of these wrenches will answer for the same purpose. The only difference I can see is that Fig. 56 may suit one brother smith and Fig. 57

another. To make the bend in Fig. 56 proceed as follows: First mark the square space on a board to the size required, then draw the lines diagonally from corner to corner as shown.

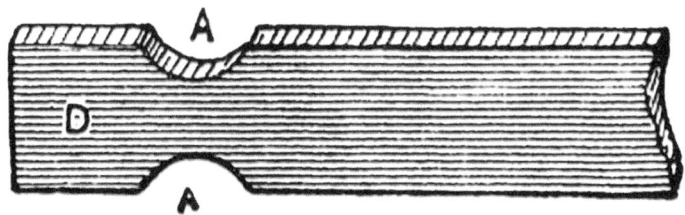

FIG. 58—SHOWS POINTS TO BE FULLERED.

This will give the center at A. Now take a compass, set it to the same length as the sides of the square, which we will find from C to D on either side.

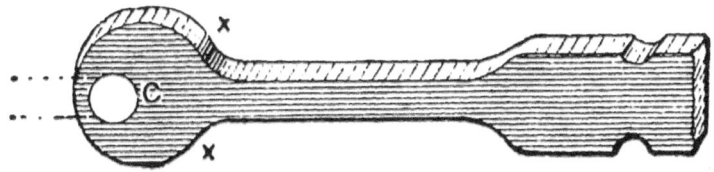

FIG. 59—SHAPING THE JAW.

Now set the compass point at E and draw the curve X to line A, then set the compass point on F and draw the curve X X to line A, which is one-half of the wrench. The point E is as far from C as the head or jaw of the wrench is from D to O.

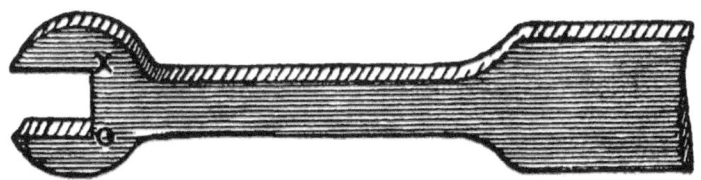

FIG. 60—SHOWS FAULTY METHOD OF FORGING JAWS.

The space between *E* and *F* is the same as the width of the wrench. With the other end of the wrench proceed the same as with the first end. By bending a wrench this way over the draft a true curve will be obtained from the center of the wrench.

Fig. 57, I think, explains itself. To make a wrench of this kind I proceed as follows: Fuller Fig. 58 at *A A*, forge and round *D* as *C* in Fig. 59; now get the center of *C* and punch the hole, letting the outside of the hole strike the center of *C*, as shown in cut. This gives us a strong corner at *X X*. Then split out as per dotted lines and finish. Never make a corner in a wrench as is the case at *X O* in Fig. 60, as it is more apt to break than when made rounding as shown in both Figs. 43 and 44. I always make these wrenches of spring or cast steel.—*By* Now & Then.

ANOTHER METHOD OF MAKING WRENCHES.

My idea of the right way to make a wrench is to get at it in the way that takes the least labor, provided equally good results are obtained. I have two steel S wrenches in use.

One is a 5-16 and ¼-inch, made from spring steel. This I have had in constant use for the last eight years. The other is a 5-16 and ⅜-inch, made from blistered steel, which I have used for the last seven years. Both are in good order still. I will try to describe my way of making a wrench of this description.

I take a piece of steel of the required dimensions and fuller it, as shown in Fig. 61.

FIG. 61—FIRST STEP IN MAKING WRENCH.

I then punch a hole, as is also shown in the same figure. I next draw out the handle, cut out from the hole, as in Fig. 62, then work up to shape and fuller.

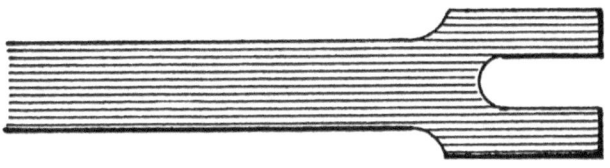

FIG. 62—HANDLE OF WRENCH DRAWN OUT.

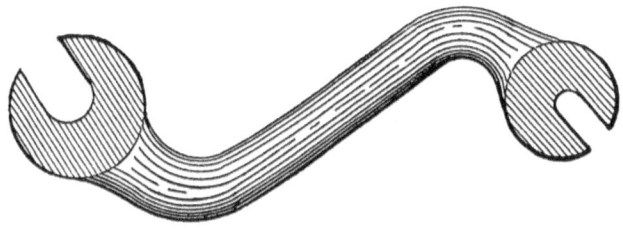

FIG. 63—THE FINISHED WRENCH.

A wrench can be made very quickly in this manner and as strong as you please. It is shown finished by Fig. 63.—*By* W. I. G

AN ADJUSTABLE WRENCH.

My method of making an adjustable wrench is as follows:

I take a piece of soft steel and forge out the jaw as shown in Fig. 64. I next take a piece of ¾-inch gas pipe, cut a thread on one end, then cut a thread in the jaw, and screw the jaw and pipe together tightly. I then heat gently, and when hot flatten down the jaw and about two-thirds of the pipe. I then heat again, having a drift key ready to fit the hole tightly.

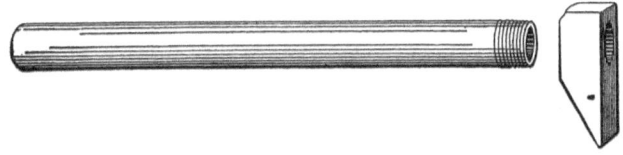

FIG. 64—SHOWING HOW THE JAW OF AN ADJUSTABLE WRENCH IS FORGED.

I drive the key in, taking care not to split the pipe by making the corners round, and then heat the other end and finish the handles to shape. I next make a ring of 1 ½ inches by 3-16-inch iron to go on the handle where the hole for the nut is to be made.

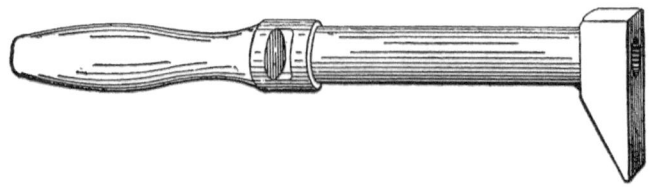

FIG. 65—SHOWING HOW THE HANDLE AND NUT ARE MADE.

The ring is heated and pressed on red hot. When it is in place and fitted down closely it will never become loose.

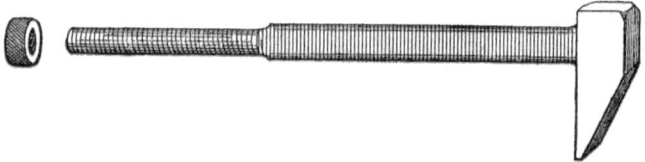

FIG. 66—SHOWING HOW THE OTHER PIECE IS FORGED.

The hole in which the nut works is made by drilling two holes close together and filing them out. The tool then looks as shown in Fig. 65. I next forge out the other part, as shown in Fig. 66, of one solid piece.

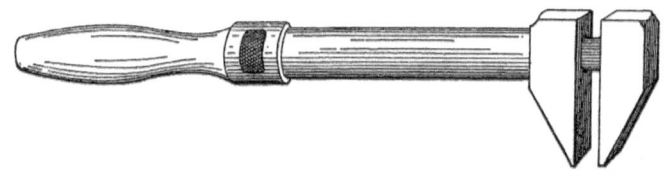

FIG. 67—THE WRENCH COMPLETED.

I cut a thread in the end, make a round nut, finish off, put together, and have the strong, neat-looking wrench shown completed in Fig. 67.

MAKING AN ADJUSTABLE WRENCH.

The accompanying engraving, Fig. 68, illustrates my way of making an adjustable 2-inch wrench, which I find very strong and handy.

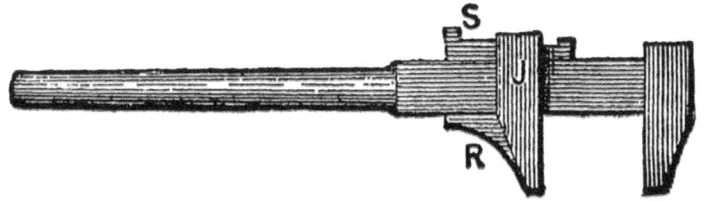

FIG. 68—MAKING AN ADJUSTABLE WRENCH

First, I forge one jaw and the handle solid in the usual way; then I forge the sliding jaw *J* with a web at *R*, and slip behind it a gib-wedge *S*, which will tighten or loosen the

FORGING A BOLT, A NUT AND A WRENCH.

I would like to give my way of forging a bolt, a nut and a wrench.

A great many blacksmiths take exception to making a bolt by welding the head on. I claim that if the bolt is upset as it should be, and the head properly welded on, it is as good, if not better, than a solid head.

The way I do such a piece of work is this: Before welding the collar on to form the head, I upset my iron to the extent that the diameter of the bolt under the head will be the same or a little larger than the original diameter of the bolt.

If this is done properly the bolt will not be weakened at all. Should you fail to upset the iron sufficiently, however, and let the collar cut into the bolt, then, of course, you will have a weak spot at the inner end of the head.

If I have a large nut to forge I do it in this way: I take iron of the required size and bend it around a mandrel, leaving the ends about an inch apart. Then, with my chisel, I cut one end down about one-half the thickness of the iron, as shown in Fig. 69.

With a small fuller I draw out the part cut down, on the horn of the anvil, so as to form a lip, as shown in Fig. 70.

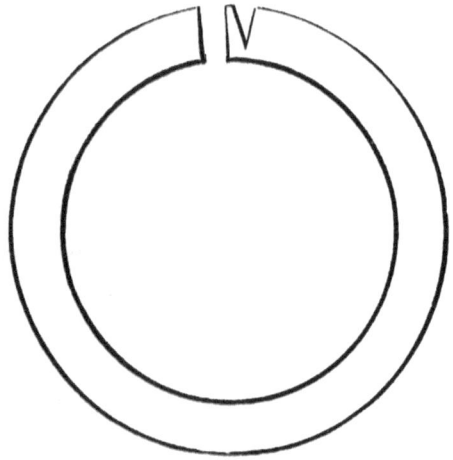

FIG. 69—PREPARING TO WELD A BOLT HEAD.

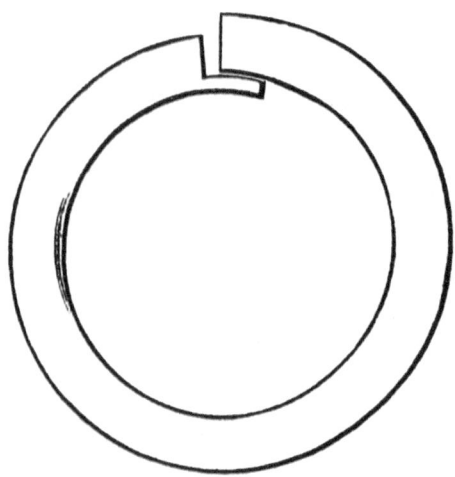

FIG. 70—FORMATION OF LIP.

Drive the ends together so that the lip described will be close to the inner side of the opposite end of the ring.

Take a good welding heat, and weld on the mandrel, and you will have a good, sound nut, with a smooth inner surface. If you undertake to forge a nut by jumping the ends together you will not make a good job of it; at least, that is my experience in such cases.

Now, in regard to a wrench. If I have a large wrench to forge, I pursue the following plan: Take a piece of iron the required width and thickness to draw out to the shape shown in Fig. 71.

Bend it as represented by Fig. 72 and scarf the part *A* as for an ordinary weld.

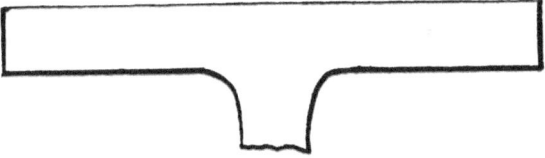

FIG. 71—FORGING A WRENCH.

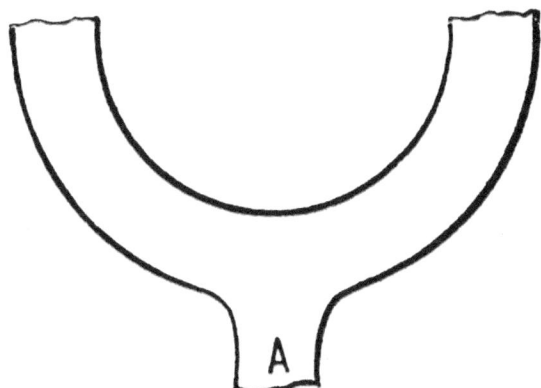

FIG. 72—BENDING PIECE TO FORM JAW.

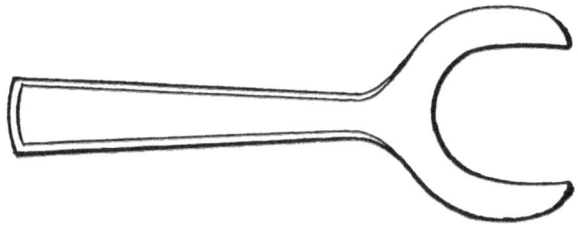

FIG. 73—THE WRENCH COMPLETE.

Then forge the handle and weld on and complete the wrench, as shown in Fig. 73.—*By* G. B. J.

CHAPTER III.

WELDING, BRAZING, SOLDERING.

The act of joining metals by the aid of heat is technically known as welding, brazing and soldering. The first is applied to iron and steel by heating the surfaces to be joined to a fusible state, then, by repeated blows, or by pressure, unite the particles and restore the whole to a condition similar to that existing before the metal was severed.

Brazing is the act of joining iron or composition metals by the use of brass heated to a fluid state in conjunction with the edges to be joined, and then allowed to cool slowly.

Soldering is the joining of metals of like or different kinds by another metal that fuses at a lower degree of heat than those which are joined, in which case the solder, only, is reduced to a fluid state. The degrees of heat, the condition of the surfaces, and the skill of the workmen, are all important factors.

The following tables will be found useful for reference:

	Degrees Fah.
The greatest heat of a smith's forge (common) is	2346
Welding heat of iron	1892

MELTING POINT OF METALS.

	Degrees Fah.
Brass	1900
Copper	1996
Lead	612
Solder (common)	475
" (plumber)	360
Tin	442
Zinc	680
Lead 1, tin 1, bismuth 4	201
Lead 2, tin 3, bismuth 5	212

SOLDERS.

Under this head is grouped compositions used for uniting metals, the proportions being by weight.

	Copper	Lead	Tin	Bismuth	Zinc	Silver	Gold	Antimony	Calcium
Tin	..	75	25	..	..	..	..	..	..
"	..	16	58	16	..	..	..	10	..
" (melts at 360°)	..	33	67	..	..	..	..	..	.
Spelter (soft)	50	.	..	..	50	..	..	..	..
" (hard)	67	..	..	..	33	..	..	..	..
Lead	..	67	33	..	..	..	..	..	..
Steel	13	..	..	..	5	82	..	..	..
Brass or copper	50	..	..	..	50	..	..	..	..
Fine brass	47	..	..	..	47	6	..	..	..
Pewter (soft)	..	45	33	22	..	..	..	..	..
" (hard)	..	25	50	25	..	..	..	..	..
Gold	4	..	..	..	..	7	89	..	..
" (hard)	66	..	..	..	34	..	..	..	..
" (soft)	..	34	66	..	..	80	..	..	..
Silver (hard)	20	..	..	..	..	67	..	..	21
Pewter	..	20	40	40	..	..	..	..	..
Iron	66	..	..	33	..	..	..	1	..
Copper	53	..	47	..	..	..	..	..	..

FUSIBLE COMPOUNDS.

	Copper	Lead	Tin	Bismuth	Zinc	Silver	Gold	Antimony	Calcium
Rose's (fusing at 200°)	..	25	25	50	..	..	..	..	..
" (fusing at less than 200°)	..	33.3	..	33.4	33.3	..	..	..	..
Fusing at 150° to 160°	..	25	12	50	..	..	..	..	20

FLUXES FOR SOLDERING OR WELDING.

Metal	Flux.
Iron	Borax
Tinned iron	Resin
Copper or brass	Sal-ammoniac
Zinc	Chloride of zinc.
Lead	Tallow or resin.
Lead and tin pipes	Resin and sweet oil.

THEORY OF WELDING.

The generally received theory of welding is that it is merely pressing the molecules of metal into contact, or, rather, into such proximity as they have in the other parts of the bar. Up to this point there can hardly be any difference of opinion, but here uncertainty begins.

What impairs or prevents welding? Is it merely the interposition of foreign substances between the molecules of iron and any other substance which will enter into molecular relations or vibrations with iron? Is it merely the mechanical preventing of contact between molecules by the interposition of such substances? This theory is based on such facts as the following:

1. Not only iron, but steel, has been so perfectly united that the seam could not be discovered, and that the strength was as great as it was at any point, by accurately planing and thoroughly smoothing and cleaning the surfaces, binding the two pieces together, subjecting them to a welding heat and pressing them together by a very few hammer blows. But when a thin film of oxide of iron was placed between similar smooth surfaces, a weld could not be effected.

2. Heterogeneous steel scrap, having a much larger variation in composition than these irons have, when placed in a box composed of wrought-iron side and end pieces laid together, is (on a commercial scale) heated to the high temperature which the wrought iron will stand, and then rolled into bars which are more homogeneous than ordinary wrought iron. The wrought iron box so settles together, as the heat increases, that it nearly excludes the oxidizing atmosphere of the furnace, and no film of oxide of iron is interposed between the surfaces. At the same time, the inclosed and more fusible steel is partially melted, so that the impurities are partly forced out and partly diffused throughout the mass by the rolling.

The other theory is that the molecular motions of the iron are changed by the presence of certain impurities, such as copper and carbon, in such a manner that welding cannot occur or is greatly impaired. In favor of this theory it may be claimed that, say, two per cent of copper will almost prevent a weld, while, if the interposition theory were true, this copper could only weaken the weld two per cent, as it could only cover two per cent of the surfaces of the molecules to be united. It is also stated that one per cent of carbon greatly impairs welding power, while the mere interposition of carbon should only reduce it one per cent.

On the other hand, it may be claimed that in the perfect welding due to the fusion of cast iron, the interposition of ten or even twenty per cent of impurities, such as carbon, silicon and copper, does not affect the strength of the mass as much as one or two per cent of carbon or copper affects the strength of a weld made at a plastic instead of a fluid heat. It is also true that high tool-steel, containing one and one-half per cent of carbon, is much stronger throughout its mass, all of which has been welded by fusion, than it would be if it had less carbon. Hence copper and carbon cannot impair the welding power of iron in any greater degree than by their interposition, provided the welding has the benefit of that *perfect mobility* which is due to fusion. The similar effect of partial fusion of steel in a wrought-iron box has already been mentioned. The inference is that imperfect welding is not the result of a change in molecular motions, due to impurities, but of imperfect mobility of the mass—of not giving the molecules a chance to get together.

Should it be suggested that the temperature of fusion, as compared with that of plasticity, may so change chemical affinities as to account for the different degrees of welding power, it may be answered that the temperature of fusion in one kind of iron is lower than that of plasticity in another, and that as the welding and melting points of iron are largely due to the carbon they contain, such an impurity as copper, for instance, ought, on this theory, to impair welding in some cases and not to affect it in others. This will be further referred to.

The next inference would be that by increasing temperature we chiefly improve the quality of welding. If temperature is increased to fusion, welding is practically perfect; if to plasticity and mobility of surfaces, welding should be nearly perfect.

Then how does it sometimes occur that the more irons are heated the worse they weld?

1. Not by reason of mere temperature; for a heat almost to dissociation will fuse wrought iron into a homogeneous mass.

2. Probably by reason of oxidation, which, in a smith's fire especially, necessarily increases as the temperature increases. Even in a gas furnace, a very hot flame is usually an oxidizing flame. The oxide of iron forms a dividing film between the surfaces to be joined, while the slight interposition of the same oxide, when diffused throughout the mass by fusion or partial fusion, hardly affects welding. It is true that the contained slag, or the artificial flux, becomes more fluid as the temperature rises, and thus tends to wash away the oxide

from the surfaces; but inasmuch as any iron, with any welding flux, can be oxidized till it scintillates, the value of a high heat in liquefying the slag is more than balanced by its damage in burning the iron.

3. But it still remains to be explained why some irons weld at a higher temperature than others; notably, why irons high in carbon or in some other impurities can only be welded soundly by ordinary processes at low heats. It can only be said that these impurities, as far as we are aware, increase the fusibility of iron, and that in an oxidizing flame oxidation becomes more excessive as the point of fusion approaches. Welding demands a certain condition of plasticity of surface; if this condition is not reached, welding fails for want of contact due to excessive oxidation. The temperature of this certain condition of plasticity varies with all the different compositions of irons. Hence, while it may be true that heterogeneous irons, which have different welding points, cannot be soundly welded to one another in an oxidizing flame, it is not yet proved, nor is it probable, that homogeneous irons cannot be welded together, whatever their composition, even in an oxidizing flame. A collateral proof of this is that one smith can weld irons and steels which another smith cannot weld at all, by means of a skillful selection of fluxes and a nice variation of temperature.

To recapitulate: It is certain that perfect welds are made by means of perfect contact, due to fusion, and that nearly perfect welds are made by means of such contact as may be got by partial fusion in a non-oxidizing atmosphere, or by the mechanical fitting of surfaces, whatever the composition of the iron may be, within all known limits. While high temperature is thus the first cause of that mobility which promotes welding, it is also the cause, in an oxidizing atmosphere, of that "burning" which injures both the weld and the iron. Hence, welding in an oxidizing atmosphere must be done at a heat which gives a compromise between imperfect contact, due to want of mobility on the one hand, and imperfect contact, due to oxidation on the other hand. This heat varies with each different composition of irons. It varies because these compositions change the fusing points of irons, and hence their points of excessive oxidation. Hence, while ingredients, such as carbon, phosphorus, copper, etc., positively do not prevent welding under fusion, or in a non-oxidizing atmosphere, it is probable that they impair it in an oxidizing atmosphere, not directly, but only by changing the susceptibility of the iron to oxidation.

The obvious conclusions are: First, That any wrought iron, of whatever ordinary composition, may be welded to itself in an oxidizing atmosphere at a certain temperature, which may differ very largely from that one which is vaguely known as a "welding heat;" second, That in a non-oxidizing atmosphere, heterogeneous irons, however impure, may be soundly welded at indefinitely high temperature. —*From the report of the United States Board appointed to test iron and steel.*

WELDS AND WELDING.

A weld to be sound, must, like everything else, be made according to sound common sense.

The theory of welding is simple enough, and only requires a little thought to make it easy to put into practice.

WELDS AND WELDING. FIG. 74—SHOWING A FAULTY METHOD OF WELDING.

If the iron is got to a proper welding heat all through its mass, there are just three things to guard against in order to get a sound job. (1) the air; (2) scale; and (3) dirt. Referring to the first, suppose the pieces are scarfed as in Fig. 74 (which is a form that beginners are very likely to make), and when the two pieces are put together, they will meet all around the edges. This simply forms a hollow pocket enclosing a certain amount of air, and also whatever amount of dirt or scale there may be upon the surfaces, and a sound weld becomes impossible. Another fault that a beginner is apt to fall into, is to make the scarf too short, as in Fig. 75, where it is seen that blows upon the top piece, *A*, will act to force it down, sliding it off the lower pieces.

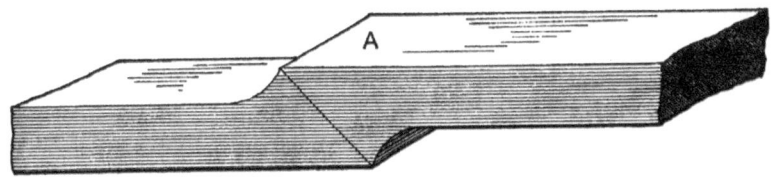

FIG. 75—SHOWING AN INSTANCE IN WHICH THE SCARF IS TOO SHORT.

We might simply lap the pieces as in Fig. 76, but the result will be an indentation in the corners C and D, and we may forge the lap down to the thickness of the base without getting this indentation out.

A similar indentation would be formed at C if lapping the end of a bar, as in Fig. 77.

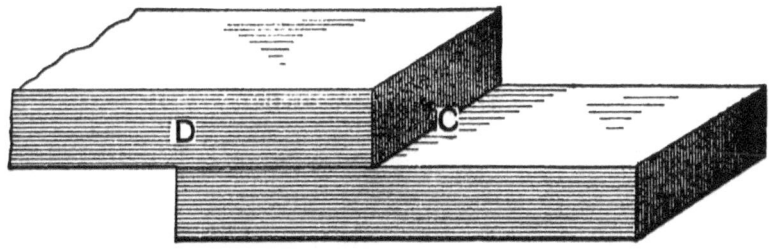

FIG. 76—SHOWING HOW THE PIECES MIGHT BE LAPPED.

Furthermore, the surfaces coming together flat are apt to enclose scale, hence such a weld can only be made either in small pieces, when the dirt has not to travel far to be worked out, or else in pieces that are heavily forged at a high heat so as to drive out the impurities.

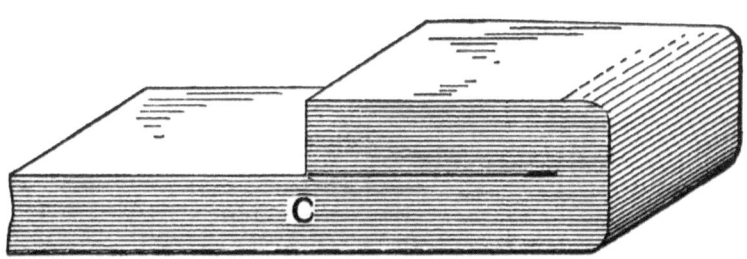

FIG. 77—SHOWING ANOTHER CASE IN WHICH AN INDENTATION WOULD BE FORMED.

The ends of two pieces, if short, may be butted as in Fig. 78, the bar being struck endwise, but this is a poor weld. In the first place only a short piece can be welded soundly in this way, because the force of the blow is lost in traveling through the weak and springy bar.

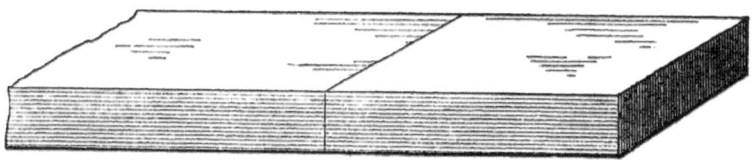

FIG. 78—SHOWING HOW THE WELD MAY BE EFFECTED BY BUTTING THE ENDS.

The secret of a sound weld (assuming of course that the iron is properly heated) lies in letting the surfaces meet at first in the middle of the weld, so that as they come together they will squeeze out the cinder, etc., and in hammering quickly.

But there are several ways of shaping the surfaces so that they will squeeze out the foreign matter: thus we may round the surfaces crosswise, as in Fig. 79, in which case a piece of scale, say at E, would be squeezed out, moving across the scarf as the surfaces were hammered together.

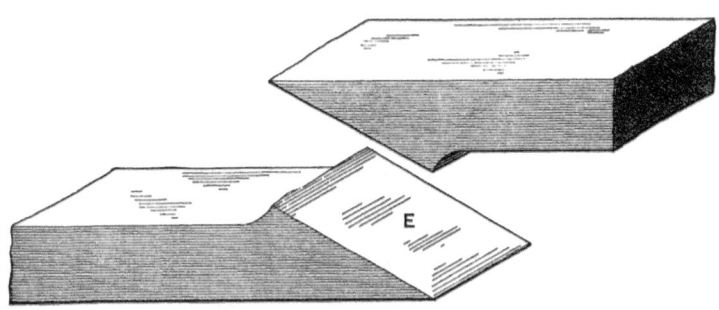

FIG. 79—SHOWING A GOOD METHOD OF SHAPING THE SURFACES.

Or we may round the scarf endwise, as in Fig. 80, but in this case the piece of foreign matter shown at F, would have to move up to G before it would be repelled.

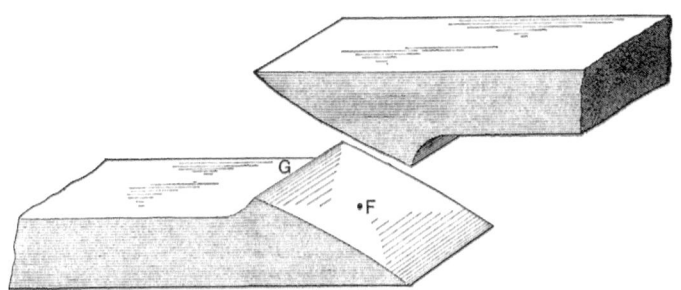

FIG. 80—SHOWING HOW THE SCARF MAY BE ROUNDED ENDWISE.

A compromise between these two plans is to round the surfaces slightly both ways, and this is the best plan all things considered.

As soon as the heat is taken from the fire, it should be quickly cleaned with a brush the instant before putting the weld together. The first hammer blows should be comparatively light and follow in quick succession.

To weld up the outer edges of the scarf and make a sightly job, a second heat should be taken if the job is large enough to require it.

An excellent example of a weld is shown in Fig. 81 where both surfaces are rounded so as to meet at *H*. In this case dirt, etc., will squeeze out sidewise as the welds come together. In all these examples the air, as well as foreign matters, is effectually excluded.

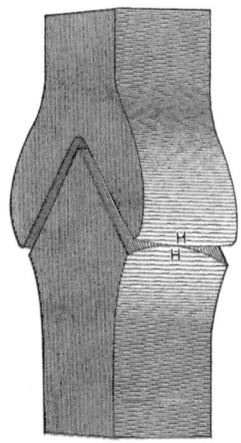

FIG. 81—SHOWING A WELD IN WHICH BOTH SURFACES ARE ROUNDED.

We now come to the welding of round bars, which are scarfed as shown in Fig. 82, so that when the two pieces are put together, as in Fig. 83, the surfaces bunch at J in the middle of the weld, and foreign matter is squeezed out all around the edges.

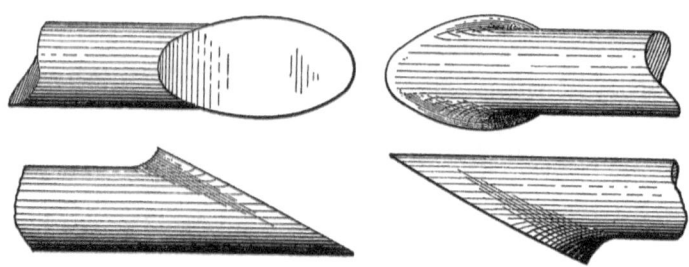

FIG. 82—SHOWING HOW ROUND BARS ARE SCARFED BEFORE BEING WELDED.

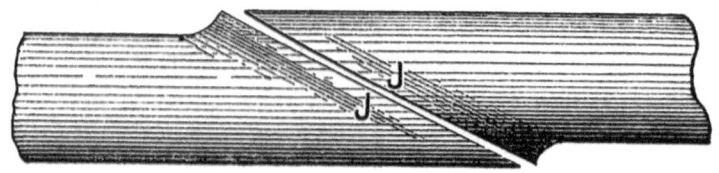

FIG. 83—SHOWING HOW THE BARS COME TOGETHER.

If the pieces to be welded are short and light, the butt weld is at least as good as any that can be made, if the ends are rounded as in Fig. 84. If the pieces are heavy and can be stood up endwise under a steam-hammer, it is still the best weld, but if the pieces are long, too much of the force of the hammer blow is lost in traveling from the end of the bar to the weld.

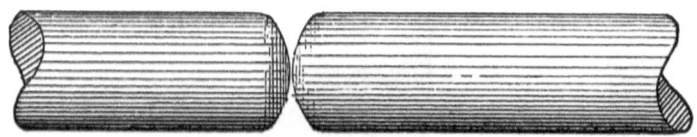

FIG. 84—SHOWING A GOOD METHOD OF WELDING WHEN THE PIECES ARE SHORT AND LIGHT.

The appearance of the weld when made and before swaging down, is shown in Fig. 85, and it is seen that the air and any dirt that may be present, is always excluded as the pieces come together.

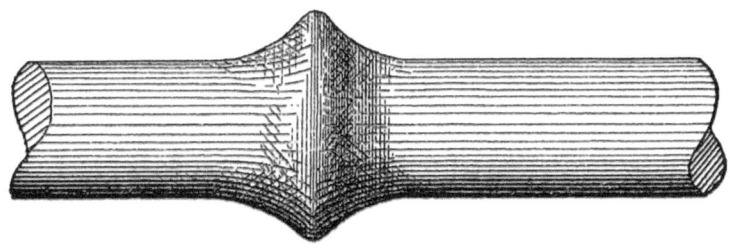

FIG. 85—SHOWING THE APPEARANCE OF THE WELD BEFORE THE SWAGING.

We now come to another class of weld where a stem is to be welded to a block, as in Fig. 86. The block is cupped as in Fig. 87, and the stem rounded and cut back as in Fig. 88, so that when the two are put together, they will meet at the point *K*, Fig. 89.

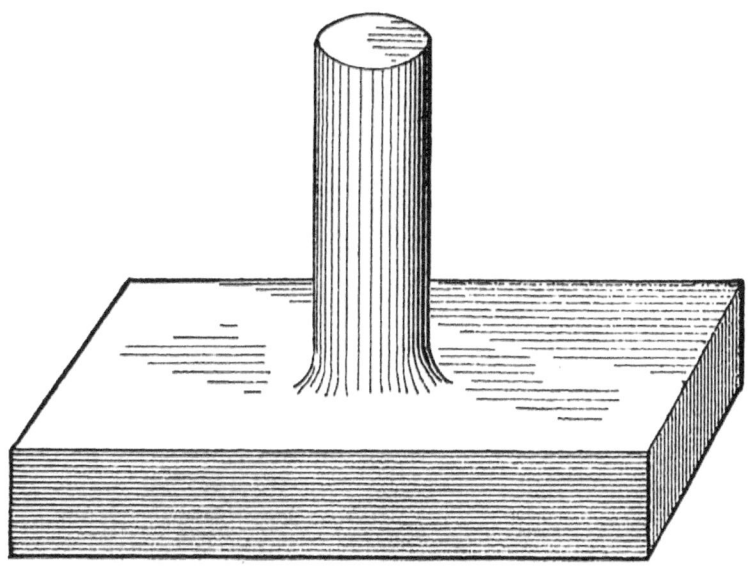

FIG. 86—SHOWING A WELD IN WHICH A STEM IS WELDED TO A BLOCK.

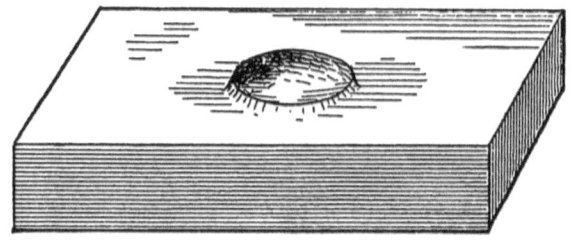

FIG. 87—SHOWING HOW THE BLOCK IS CUPPED.

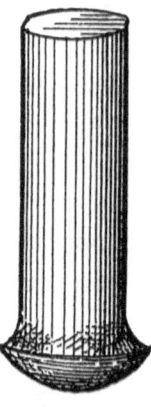

FIG. 88—SHOWING HOW THE STEM IS SHAPED.

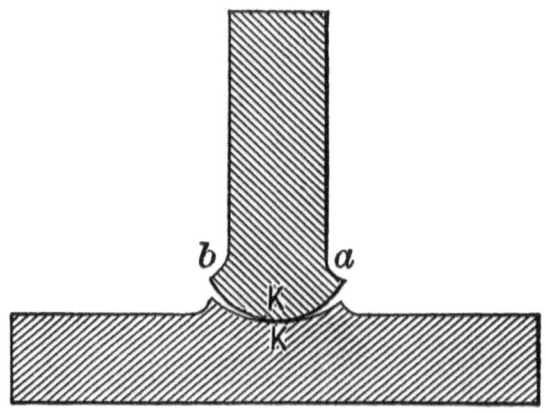

FIG. 89—SHOWING HOW THE STEM AND BLOCK ARE PUT TOGETHER.

The dirt and air will be forced upwards and outwards in this case. If the stem is short it may be driven home on the end, and fullered afterwards at the shoulders a and b, but if long the fullering only can be used to make the weld, and a good shoulder at *a, b,* is necessary.

In the days when blacksmiths made their own swages and fullers (and this is done in most first-class blacksmith shops at the present day in England), the heads of swages and fullers were made by rolling up a band of iron, as in Fig. 90.

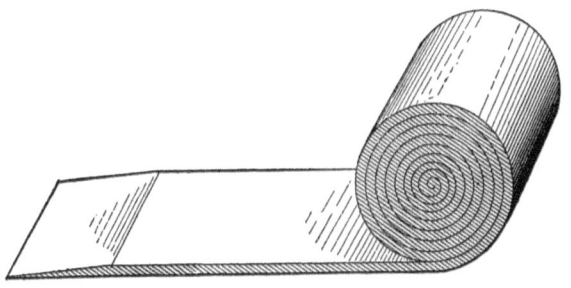

FIG. 90—SHOWING HOW THE HEADS OF SWAGES AND FULLERS ARE MADE.

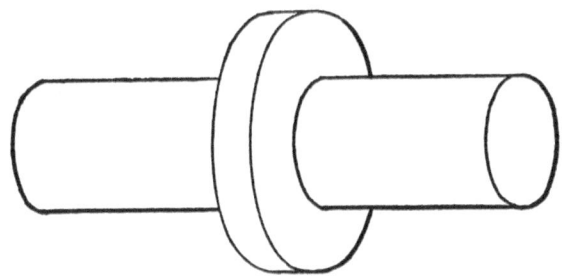

FIG. 91—SHOWING A COLLAR WELDED ON A STEM.

In this case the first hammering must be given to the outside and not to the ends of the roll, the end of the band being turned down so that it will roll down in the center. With good iron and first-class workmanship, this makes a good tool.

An example of welding a collar on a stem is shown in Figs. 91, 92, 93 and 94. Fig. 91 is the finished iron; Fig. 93 the stem jumped up in the middle to receive the collar; Fig. 92 the collar ready to be cut off the bar; and Fig. 94 the collar placed on the stem, ready for the welding heat.

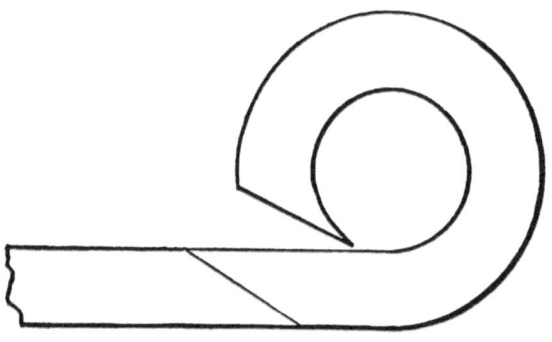

FIG. 92—THE COLLAR READY TO BE CUT.

FIG. 93—THE STEM PREPARED TO RECEIVE THE COLLAR.

Unless the stem is jumped up as shown, and the collar well beaded on it, there will be a depression or crack at the corners. Very thin washers are welded with the scarf made, as in Fig. 94, and made to overlap well.

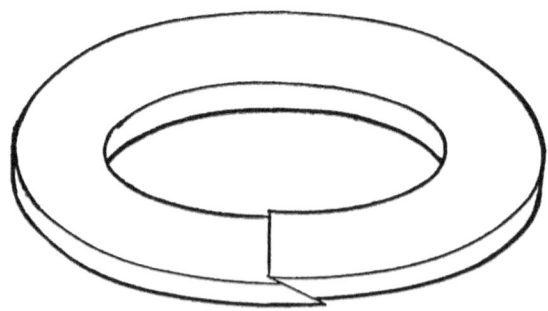

FIG. 94—THE COLLAR READY FOR THE WELDING HEAT.

In any weld, dispatch and decision are necessary elements as soon as the heat has left the fire, the thinking being mainly done while the heat is in the fire.—*By* Joshua Rose.

WELDING IRON AND STEEL.

A series of experiments were undertaken by Prof. J. Bauschinger at the instance of an engineering firm. Similar experiments had been previously made at the Royal Technical Experimental Institute, at Berlin, and by Mr. W. Hupfield, at Prevali, which gave very different results, those at Berlin being very unfavorable, those at Prevali very favorable, as regarded the welding capacity of steel. Prof. Bauschinger recapitulates the main results of these tests before describing those made by himself. The test pieces were flat, round and square in sections, the largest being 3.149 by 1.181 inches. Each piece was swelled up on the anvil, when hot, 0.196 to 0.392 inch, and after heating to the proper degree, the two pieces were laid on each other and welded together by hand or steam hammer.

In the chief experiment the steam hammer was employed. Every piece after welding was tested in the usual way for tensile strength, the limit of elasticity, contraction, extension and ultimate strength being determined, the same quantities having been measured for pieces of exactly similar quality, section and length, but without a weld. The limit of elasticity in both steel and iron is nearly always reduced by welding, and this is, without exception, the case as regards the extension; the contraction of welded is less than that of unwelded pieces when the fracture takes place in the welded portion. The general conclusions arrived at are that for steel the best welding temperature is just at the transition from a red to a white heat; a quick fire and smart handling are necessary, as the pieces should not be long in the fire.—*Midland Industrial Gazette.*

POINTS ABOUT WELDING.

To obtain a good sound weld, the following points should be observed:

The scarf should be sufficiently larger than the finished size to permit the weld to be full size after welding. The joint surface of the scarf should be slightly rounding, so that, when the two pieces are placed together to weld, there will be no air inclosed between them.

They should be heated in a clear fire of bright and not gaseous coal. Thick pieces should not be heated too quickly, or the interior metal will not be brought up to the required heat. They should be frequently turned in the fire, to insure uniformity of temperature, and be made as hot as possible without burning them.

They should be withdrawn from the fire occasionally and sprinkled with sand, which serves to exclude the air from the surface and prevent oxidation, and at the same time cools the outer surface and thin edges, giving the interior metal and thicker parts time to become heated all through.

When the pieces are placed upon the anvil to weld them, they should be quickly cleaned with either a wire brush or a piece of wood made ragged by having been hammered. The scarfs should be placed to well overlap each other, and should receive light and quickly succeeding blows at first, and heavier ones afterward.

As soon as the pieces are firmly joined, the hammer blows should be delivered with a view to close the edges of the scarf, so that the joint of the weld shall not show when the job is finished.

WELDING CAST STEEL FORKS.

I desire to say with regard to springs, cast steel forks and other similar articles of this general kind, also with regard to spring-tempering in a country job shop, that I have been troubled in the same manner as other smiths. I have tried the same remedies that they have tried.

When I learned my trade I had occasion to mend forks, and had experience on other difficult jobs of the same kind. Not knowing how to hold the parts until I could weld them, I commenced by scarfing and punching, and then welding, also by riveting the parts together. This was not satisfactory, as they frequently broke at the riveting holes. I tried every device that I could think of, splitting and locking them together, sometimes putting in a good piece of iron or steel as the occasion required. At last I tried scarfing and lapping the ends together, and holding them together with forge tongs at one end of the lap until I could get a light borax heat to fasten the other end of the lap together. Then, by taking another good heat and welding the whole together, and drawing to their proper size and shape, I obtained a satisfactory job. Of late years I have found it a very great help in welding to keep some clean filings, and to use them between the laps. The filings cause the parts to unite very much more readily.—*By* A. H.

WELDING STEEL.

I have seen men try to weld steel in a fire where it would be impossible to weld iron; they prepared the pieces for welding skillfully, but they did not use borax in the best manner. They used it at times too freely, at other times not enough, using at the wrong time and not applying it on the right place. I will give my way of welding steel:

See that your fire is clean from all cinders and ashes, then take selected coal and build a fire so large that you will not have to add any unburnt coal while welding. Then prepare the steel which you wish to weld by upsetting both pieces near the ends, scarfing carefully, and when you can do it, punch a hole and rivet them together. Let the lap be from half an inch to an inch, according to the size of steel you wish to weld, and have the lap fit as snug as possible all around. Place the steel in the fire and heat to a low cherry, then apply borax to the part which is to be heated. Apply the borax not only on the lap but also next to the lap, but do not use too much. Then bring to a welding heat and strike quickly with light blows.—*By* G. K.

WELDING STEEL TIRES.

Bessemer steel tires may be welded almost as readily as iron with the ordinary borax flux. Crucible steel tires require a little more coal and a lower heat.—*By* Old Tire.

WELDING TIRES.

No. 1. My plan in welding small tires, which works well, is to put good iron filings between the scarfs and avoid heating hot enough to burn. This plan will work fully as well on old tires as on new, and especially when you do not happen to get a weld the first heat.—*By* H. A. S.

No. 2. Open the unwelded lap of your tire and insert (if your tire is steel) steel filings (if iron, iron filings); close the lap, add your flux and weld at a fluxing heat.—*By* Tire Setter.

No. 3. Our way of welding tires is to cut the bar three times and three-fourths the thickness of the iron longer than the wheel measures. We then upset it thoroughly enough to get a good heavy scarf. In welding, instead of laying one end on top of the other, we put the ends of the scarf band together, place them in the fire, and bring them to a nearly white heat, then put them on the anvil and lap the welds, then sand and put in the fire. By so doing we have the top lap hot and get a weld thoroughly, without burning off any at the bottom in so doing. We get a nice, smooth weld, and the outside corners are flush and full, and show no canker spots on the tire. This method is specially for heavy tires, but it is a good plan for all sizes of tire.—*By* B. & S.

DO NOT BURN YOUR TIRES IN WELDING.

I would like to call the attention of carriage smiths to a great evil that many fall into when welding tires, viz.: of allowing the tire to burn on each side of the lap while taking a heat.

Many smiths fail to take into consideration the fact that it is impossible to heat a piece of iron two inches thick, especially when it is formed of two pieces of equal thickness, one placed upon the other, as quickly as one of half the thickness could be heated, and hence, having lapped their tire, the full force of the blast is thrown upon it.

As a result, the tire is put into service with a weakness at each side of the weld, caused by being burnt while the weld was being brought to the required heat. That there is not a particle of need for such carelessness every smith knows, no matter how poor a workman he may be.

Give your weld a gradual heat; attend to it yourself and not throw the responsibility upon your helper. Have a clean lot of coal and under all a clean fire, and you will never lose a good customer by having him discover a rotten place in his tire, causing it to break when far from a forge.—*By* J. P. B.

WELDING AXLES.

I will try to describe the way we weld axles in our shop. We first get the length between the collars and then cut them off, allowing on each piece three-fourths of an inch on each back axle for waste in welding. If we wish to make a hole in the front axle, the piece is made one and one-half inches longer. The piece is then heated to near a white heat and the end is pounded down on the anvil, until an end is made which is of good size, and also as flat as possible. Notches about half an inch apart, and a quarter of an inch deep are then made in the end with a chisel. The two pieces are then put in the fire with the ends together, and when they get to a welding heat, one man takes one piece, a second man takes the other, and the ends are put together true, and one of the men strikes a few blows on them with a wooden maul. Then the joint is hammered with the pene of a small hammer, set in a long handle. The piece is kept in the fire, with the bellows blowing all the time, so as to get a good welding heat. Finally five or six heavy blows are given with the maul, and the piece is then taken out and hammered on the anvil to the size desired. If it is a nut axle, the nuts should be put on to avoid battering the threads.—*By* J. K.

WELDING CAST IRON.

The question is often asked, Can cast iron be welded to wrought iron? I will give you what I call a practical job: To weld cast iron sleigh shoes that have been broken when not worn out, I take the ends that I wish to weld together, cover them with borax, heat them to a nice mellow beat, lay them on a plain table of iron, so that when put together they will be straight on the runner. One of the pieces is held by a pin at the end; then I press the other end against it sufficiently to upset it a little, rubbing it with the face of my hammer until it is smooth. Allow it to cool and the job is done. I broke one in pieces and put it together, marking the welds with a prick punch, so as to know where it broke. It has been on a sled carrying heavy logs two and a half months, and has stood well on bare ground.—*By* Frank E. Niles.

Cast iron can be welded by heating it nearly to a melting point in a clear fire, free from dirt, and hammering very lightly. But this job requires practice and great care. A plow point can be made as hard as glass by heating nearly to a welding degree, then having a piece of cast iron hot enough to run over the joint and finally putting it in the slack tub.—*By* M. T.

WELDING MALLEABLE IRON.

Malleable cast iron may be welded together, or welded to steel or iron by the same process as you would weld two pieces of steel. Experiment first with two useless pieces. A few attempts will enable you to become an expert at the business.—*By* H. S.

WELDING MALLEABLE CAST IRON PLATES.

You can weld malleable cast iron plates by riveting them together and using a flux of powdered borax and Norwegian or crucible steel filings, equal parts. Let the first blows with your hammer be tender ones.—*By* Dandy.

WELDING CAST AND WROUGHT IRON.

It is no trick, but is easily done, if you know how. When a cast iron point is worn out, I break it off square and weld on another from one-quarter to one-half pound weight, making the point as good, and many say better than it was when new, from the fact that it sticks to the ground better.—*By* A. D.

WELDING STEEL TO A CASt IRON PLOW POINT.

My experience in welding cast steel to a cast iron plowshare has not been very much, although I succeeded in welding the first one I tried. My plan is first to heat the metal hot enough so that you can spall it off (on the end you wish to weld to) as square as possible. Then make the steel point square also, to fit the metal as neatly as possible. Take heats on both with borax, heating the steel as high as possible without burning it, and the metal also as hot as can be heated without crumbling. Then jump them together, applying the hammer smartly, but not too hard, on the steel end for several seconds. When you see the heat getting off, stop hammering, and lay the job away where it will not be disturbed until perfectly cool. You may then heat the point and sharpen or dress it to suit yourself. Do not strike on the weld, as you will knock it loose. Let it wear smooth. I do not exactly call this welding, but rather cementing the parts together, which I think is the only way that it can be done.

WELDING PLOW LAYS TO LANDSIDES.

I have found a good method for welding steel plow lays to landsides. It is as follows: After heating to a good bright red, put on plenty of wax, and when the wax is melted, put on dust from the anvil block, then take a good mellow heat and keep the top part of the lay from burning by throwing on sand. Use a light hammer and a stiff pair of tongs large enough to squeeze the lay and landside together and hold them solid while using the hammer. My object in using a light hammer, is to enable me to strike quicker and light blows, such blows being less likely to make the steel fly to pieces.—*By* C. N. Lion.

TO WELD CAST STEEL.

Take rock saltpetre, one-quarter pound, dissolve in one-quarter pound oil of vitriol, and add it to one gallon of water. After scraping the steel, get it hot and quench in the liquid, and then heat, and it will weld like iron. Better than borax.—*By* A. D. S.

TO WELD STEEL PLATE TO IRON PLATE.

To weld a steel plate to an iron plate, I would say, for a common fire you want a stout porter bar on your iron plate first—something to hold on to. If you have much of that work to do, you want a grate-fire and to use anthracite coal on the grate, with soft coal on top. A grate gives more heating area and uniformity of heat in the fire surrounding the work. The opening to the fire can be made with a piece of iron about fourteen inches long, bent at each end about three inches, and arched in the center. Place this in front and fill the center with cut wood, and cover all with soft coal; let the wood burn out, and you have a good fire with a clear entrance, like a furnace. Feed the fire till mellow, and insert the iron and steel together till red-hot, Have fluxing matter in a pepper-pot, and dust over the surface to be welded. Let it vitrify. Now place the steel and iron together, draw heat, flux at pleasure with a spoon five feet long. When the spoon sticks, the heat is nearly up. Draw, when ready. A single "drop" blow will weld, or ten blows with a heavy sledge. Cut off and finish. A charcoal, coke, or anthracite muffle-furnace is a fine thing for this work. If you are a good mechanic this will help you out. If inexperienced, you will be apt to get the fidgets, and had better let the job out.—*By* S. C.

A PRACTICAL METHOD OF WELDING BROKEN SPRING PLATES.

First heat, upset and scarf the broken ends of the "plate" (leaf we generally call it), just the same as if it were iron, and to be welded by separate sand heats; then place them in a clean fire and take separate borax heats; heat as high as the safety of the steel will permit. I then let my helper take out one end while I take the other, and when "stuck" come on to it lively with the sledge, and nine times out of ten I get a good sound weld at the first heat. Care must be taken to leave it as "heavy" at the weld as the original size of the steel.

The loss of length can generally be made good—if a "bed leaf"—by letting out, or taking up (as the case requires) the scroll at the end of the bottom "bed leaf," or, if it is an outside or intermediate leaf, the little loss in length is of no importance.

Now the welded end of the "leaf" must be tempered —a spring temper— or, no matter how perfect the weld, it will not prove a good job. Tempering properly is the most important feature of the job, and a majority of smiths overlook it altogether.

My plan is (and it is the most satisfactory and practical that I have heard of), after hardening, draw to temper by passing forth and back over a clear fire until it will ignite a pine stick when rubbed over the surface. If it is high steel, heat until the pine will blaze when rubbed hard. If it is low steel, let it sparkle only; lay down to cool.

Since I have learned through my own observation the importance of tempering, I seldom weld a spring which fails to stand as well as its fellows.—*By* Hand Hammer.

WELDING BUGGY SPRINGS.

A good way of welding buggy springs is to first scarf the broken ends down to a sharp edge, and then split them back three-quarters of an inch. Then turn one point up, the other down, as shown by Fig. 95, having ready two thin strips of iron made pretty wide in the middle, and nearly to a point at the ends. Drive the ends together, and then drive the strips of iron in between the laps from each edge, as shown in the accompanying illustration. The inside edges of the strips of iron should be made thin and sharp.

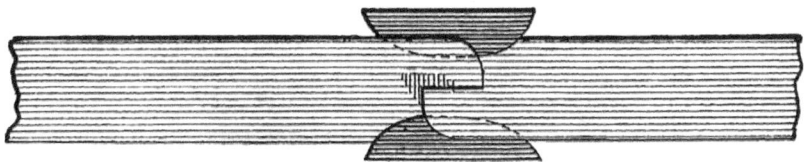

FIG. 95—WELDING A BUGGY SPRING, AS DONE BY JOHN ZECK.

I have welded springs in this way for many years, and it has never failed. The iron insures a perfect weld, and also makes up for the waste in working.—*By* John Zeck.

WELDING SPRINGS.

No. 1. It is a fact that steel will waste away some by heating at a welding heat, and if it wastes away it will be thinner at each side of the weld than elsewhere, and more liable to break at the weak point, and more especially after taking two welding heats to one weld, and if not upset it must necessarily leave an imperfect job.

I hardly see the necessity of taking more than one heat on a weld, provided it is well executed. I will give you my method, I first upset each piece of steel, weld and chamfer nicely; then punch a small hole in each piece for a rivet. I then, at the lap, slip between the two pieces a thin piece of steel with a rivet hole in it, and rivet the three together. Each end of the middle piece is hammered sharp, and the amount must be about what will waste away in welding. Put each end of the weld alternately in the vise and hammer down the opposite end so that it lies close to the bar, then borax and weld down with one heat. Let the helper come down lively. It is not necessary to let the under point of lap come in contact with the anvil while welding, but it may extend beyond till a few blows are struck on the opposite end, and the reason is that the anvil would, perhaps, cool the thin part below the welding point.

The main object is to get a good weld, and one heat is far better than more for this purpose. Perhaps I had better add, make as short a lap as possible.— By Frank.

No. 2. I don't believe that a substantial weld can be made by the use of rivets.

I upset each piece of steel one inch back from the end, leaving the end its natural size. I then split the end twice and chamfer nicely; turn the center point of the split end down and the two outside points up. After preparing both pieces in this way, I slip them together and fit them both edgewise and flatwise, forming a lock and making the lap as short as possible. I try to make my weld at one heat.—By A. F.

No. 3. I will state how I have had the best results. Heat to a cherry red, chamfer the ends short, split into three equal parts about five-eighths up. Bend the two outsides one way and the center the other way of both pieces to be welded. Then put together and close the laps down, making them so they will stick together; then take a light borax heat and work quick. Do not hammer too much on one side, but turn in quick succession to prevent from chilling the scarf. This way of lapping springs requires no upsetting of the ends, and is the best way I ever saw to put springs together to get a heat. Hammer all the temper in that you give them.—By G. W. W.

No. 4. My way of welding springs is as follows:

I first upset the ends, then draw them down to a point, as shown by Fig. 96, then punch, rivet, and weld in a fire, taking care to first burn all gas out of the coal. I apply borax freely, heat slowly and make the hammer blows quick and hard.

FIG. 96—WELDING SPRINGS AS DONE BY "C. D."

To ensure success I usually take two welding heats. There will be no thin edges in the laps, and if care is taken to avoid burning the steel, the job will be one that can be warranted every time.—*By* C. D.

No. 5. My plan for welding springs is to upset to about the original thickness. Make a neat short scarf and weld the same as two pieces of common iron. I never split or punch holes. I find I can get a far better job otherwise. If, by accident, I should not get a good job, and some part is inclined to scale up and not weld thoroughly, I take a small piece of rusty hoop-iron and insert under the scale and take a second heat. There is a good deal in having a good, clean, well-charred fire in work of this kind.—*By* A. M. B.

No. 6. Will you permit me to say a few words about welding springs? Some years ago I was told by Mr. Frank Wright, of Waltham, Mass., to try one and pin or rivet it. I did so, and have employed that method ever since. If the end is split twice and the pieces locked together it is impossible to get a good solid weld, as it is necessary to weld the ends down, three on each side, and to weld the split lengthwise. Several years since I welded the main and second leaf of a 1 ¾-inch 5-leaf spring for the hind end of a wagon that was calculated to carry from 1,200 to 1,700 pounds, three trips per week. I know that it was used for three years after the time I repaired it, and it never broke or settled. By riveting the pieces together I find I can weld seven out of every ten and do the work so perfect that the line of weld cannot be discovered.—*By* L.

No. 7. My plan is to scarf each end and to punch a hole in about a half inch from the end, as at *A*, Fig. 97, then get a thin piece of iron and punch a hole in it, and lay it between the laps and put a rivet through it and the laps as at *C*, and then weld with borax. By welding in this way I never met with a failure, and I consider it better than splitting the ends.—*By* W. B.

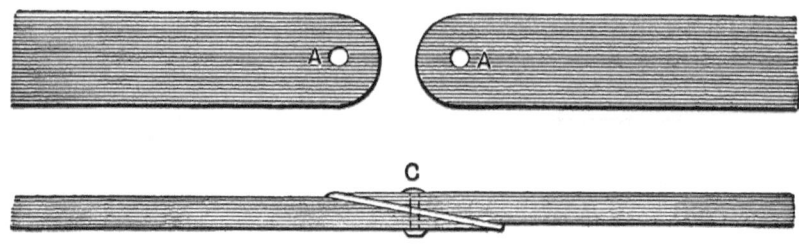

FIG. 97—WELDING SPRINGS BY THE METHOD OF "W. B."

No. 8. I will tell you my way of welding springs. I do not upset the pieces, but chamfer them off fine, and punch two holes in each end, giving them about one and a half or two inches lap; rivet them firmly together, to prevent any possible displacement. Then I give them as low a borax heat as practicable to make a weld on one side near the chamfer of one end, and follow by a like operation on the other side; the drawing of the weld to the natural thickness and breadth of the leaf will fetch it also to its proper length.

The loss of the metal in such a weld is not noticeable. I seldom ever have any springs come back after such treatment. My objection to heating the pieces separately is that it requires too strong heat to make them stick. Though the steel may not be actually burnt, it will lose much of its strength by overheating, and will be very liable to break at the same place again, though the weld may be perfect. Besides, it requires more quick and accurate movements to bring the pieces together on the anvil. I give no other temper than a gentle hammering, which I keep up a little while after the leaf is black.— By O. D.

WELDING AND TEMPERING SPRINGS—MENDING A SPRING WITH A BROKEN EAR.

My way of welding springs may be of some interest. I first upset, then scarf a little, next split, care being taken not to split too deep, or you are liable to get too long a lap. Next turn the ends each way and scarf with the ball of the hammer, letting them flatten towards the center of the spring. Next warm them up, let the helper take one end and then press together. Beat down the laps and the welding will be almost as strong as if rivets had been used, without the disadvantage of rivets. In heating to weld, most of us make the mistake of heating too fast. The best results are got by heating as slowly as possible. I want the helper in welding to strike quickly and not too hard, as heavy blows are apt to drive the laps over to one side.

In tempering I take a light hammer, and hammer the spring until it is nearly cold. A spring treated in this manner will stand as much as any weld.

To repair a spring with the ears broken off, close up. I upset it, then take a piece of good iron, scarf the edges, cut half way through it, lap the ends over the end of the spring, weld it up in good shape, turn the ears on a bottom fuller; finish the end; and after filing it up, it cannot be distinguished from the other without close inspection.—*By* Observer.

HOW TO MEND WAGON SPRInGS.

To mend a wagon spring so that it will be as strong and durable as before it was broken is no easy job, and nine-tenths of the smiths will tell you it cannot be done. I am of the opinion that if the work is carefully and properly done, nineteen-twentieths of the springs mended will do good service. There are various ways of mending a spring, and every smith has a way of doing the work peculiar to himself, and, of course, thinks his method is the best.

I first upset the pieces to be mended to such a thickness that when the work is done, there shall be no waste near the weld, the place where mended springs usually break. I next carry the pieces about one inch back, punch and put in a small rivet to keep them in place while taking a welding heat. After obtaining a good borax heat I make the weld. If it is not perfect I take a second or a third borax heat until a solid and uniform weld is obtained. Much depends upon uniformity of the weld. If the spring is left thickest at the weld a break near it is liable to occur, especially in the case of an inside leaf. The weld should be so perfect and uniform that it cannot be easily seen where the spring was welded. I never attempt to put any temper into a spring, any more than what can be done by hammering.—*By* W. H. B.

WELDING SPRINGS.

My way of welding a spring is as follows: I chamfer the ends and split them, as shown by Fig. 98, then place them together as closely and as firmly as possible. I then give the spring a bright red heat, next roll it in calcined borax, then place it in a clean fire of coal well coked, and blow very lightly, frequently baring the splice to see that the heat is strictly in the right place and to also make sure that it does not burn. The blower can hardly be worked too slowly, for the steel must have time to heat all through alike. It will take longer when the blower crank is moved so slowly, but the time is not wasted.

FIG. 98—WELDING SPRINGS AS DONE BY A "JERSEY BLACKSMITH."

Near the end of the heat I keep the steel bared nearly all the time, watching it closely and just before it comes to a sparkle I cover it and give it three or four good heavy blasts, then take it out and strike as rapidly as possible upon the several lips or splices. Before drawing it down to its usual size I see that all the welds are perfect, and if they are not I go over the work again paying particular attention to the unwelded parts. I hammer until it is black for the temper. No spring I ever welded in this way has been broken, and I warrant everyone if the steel is good.—*By* Jersey Blacksmith.

HOW TO WELD
A BUGGY SPRING.

I will give my way of welding a spring which I learned from a tramp six years ago, have used ever since, and have never had a spring break at the weld. I first lay the broken ends of the springs together to get the length, which I take with the dividers set to small center punch marks in each piece. I then place the two ends in a clean fire and hammer back a short scarf, thus giving the necessary upset.

I next punch a small hole close to the ends of each piece to receive the rivet that holds them together while welding.

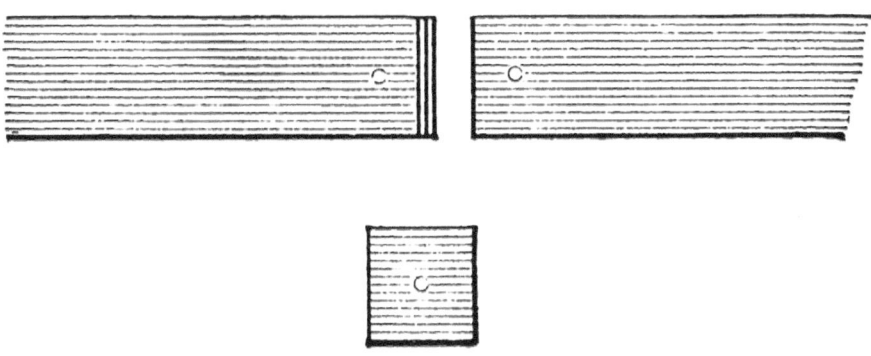

FIG. 99- SHOWS HOW "D. F. K." WELDS A BUGGY SPRING.

I next take a piece of good Norway iron, flatten to one-sixteenth of an inch in thickness, the width of the spring, and just long enough to pass the laps When placed between them, punch a hole in this for the rivet, place this piece between the laps and rivet them together; heat enough to melt borax. Lift out of the fire and soak well with borax (for you can't use too much on this kind of weld), replace in fire and heat carefully. Weld with rapid light blows, sticking both sides safely before striking on the edge; finish with as few heats as possible to make a solid job. Allow it to cool slowly, as it will require no other tempering. Fig. 99 shows how the spring and piece of Norway iron are prepared for welding. By following these directions you can weld a spring that you can safely warrant.—*By* D. F. K.

WELDING SHAFT IRONS FOR BUGGIES.

My method of welding such work as shaft irons for buggies is as follows: Fig. 100 shows the two pieces prepared for the welding process, and Fig. 101 shows the finished iron.

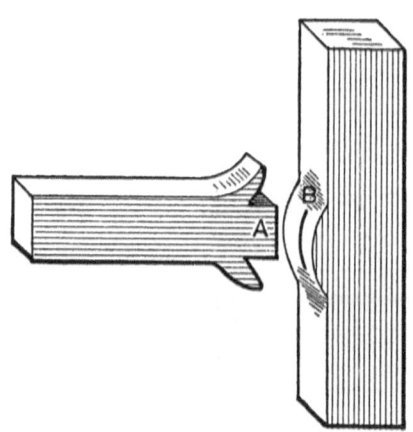

WELDING SHAFT IRONS FOR BUGGIES. FIG. 100—SHOWING THE TWO PIECES PREPARED FOR WELDING.

The projection on Fig. 101 is split to receive the tongue *A* on Fig. 100.

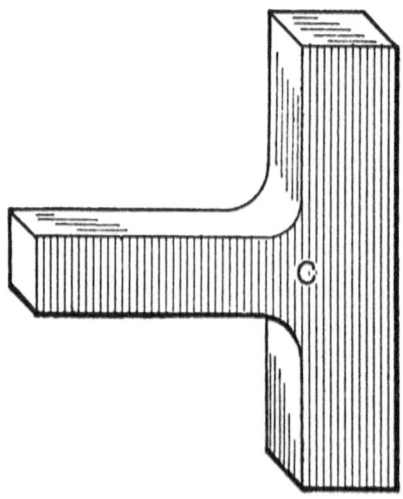

FIG. 101—THE FINISHED IRON.

I get the rounded corners at *C* (where the strain comes) by widening at *A*, and by the projection at *B*. Fig. 102 shows the old way of forming the weld, the fault in this being that there are no rounded corners as at *C* in Fig. 101, and therefore the iron is apt to break in the neck.—*By* Southern Blacksmith.

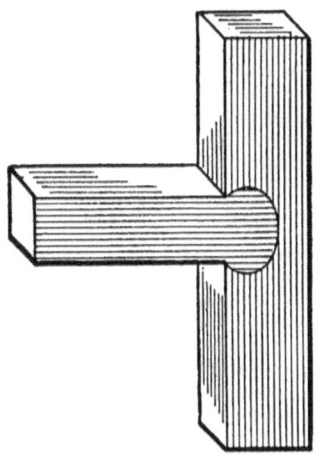

FIG. 102—SHOWING THE OLD WAY OF FORMING THE WELD.

WELDING A COLLAR ON ROUND IRON.

My method of welding a collar on round iron is as follows: Suppose the bar to be one inch in diameter.

The seat of the collar must be jumped one-eighth of an inch larger, as at *A* in Fig. 103. Leave the collar *C* open one-quarter inch, and while the washer is cold and the bar red-hot, swage the collar on so that it will hold. Then take a welding heat. If the bar is not jumped the neck will shrink in as at *B* in Fig. 104, and will probably show a crack there.—*By* J. R.

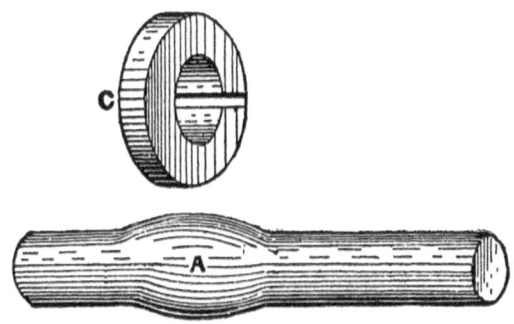

WELDING A COLLAR ON ROUND IRON, AS DONE BY "J. R." FIG. 103— SHOWING THE BAR JUMPED AT A, AND THE COLLAR C.

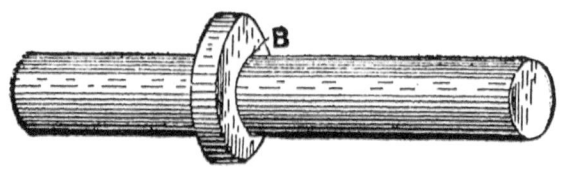

FIG. 104—SHOWING THE POINT B, WHERE THE NECK WILL SHRINK IF THE BAR IS NOT JUMPED.

WELDING A ROUND SHAFT.

I will describe my method of welding a round shaft. *A* and *B*, Fig. 105, represent the two pieces to be welded. I make one end of one piece the shape of a dot punch and one end of the other piece is counter-sunk to fit the punch shaped piece.

I then place them in the fire together, as shown in the illustration, and weld them in the fire. After they are welded I take a good welding heat, lay the piece on the anvil and work the part smooth.

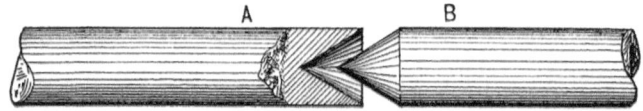

FIG. 105—WELDING A ROUND SHAFT, BY THE METHOD OF "H. H. L."

This plan insures a good weld every time, no matter what may be the size or length of the rod. In welding it be sure to place it in the position shown in the illustration, and strike at both ends while heating.—*By* H. H. L.

WELDING.

Here are some examples of heavy welding, such as jump welding long shafts, and scarf and swallow fork welds for the same.

In my opinion the jump weld is decidedly the best, if properly done, as it can be bent after welding to a right angle without showing any sign of the weld.

FIG. 106.

To make a jump weld properly the ends should be rounded, as shown in Fig. 106, and, after being brought to a welding heat with a good blast, drive them together with very light hammer blows, as a heavy blow would cause them to bounce.

As the welding proceeds employ heavier blows and the weld will drive up as in Fig. 107, after which the weld may be forged and swaged down.

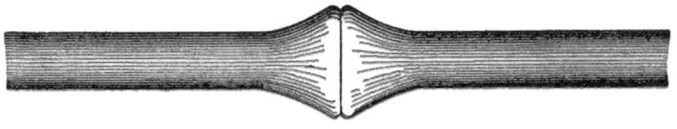

FIG. 107.

This weld will bend to a right angle, as shown in Fig. 108, without showing the scarf, whereas a scarf weld, such as shown in Fig. 109, would show, when bent, its scarf, as at *A*, in Fig. 110, while a swallow fork weld, such as is shown in Fig. 111, when bent will show its scarf, as in Fig. 112. Only the jump weld, you will see, will stand such bending and show a sound and complete weld.— *By* Southern Blacksmith.

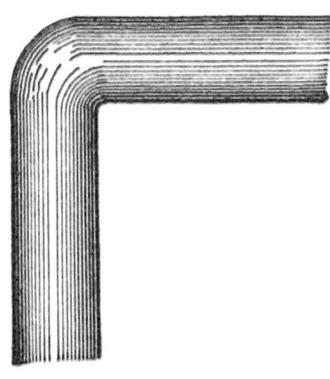

FIG. 108.

FIG. 109.

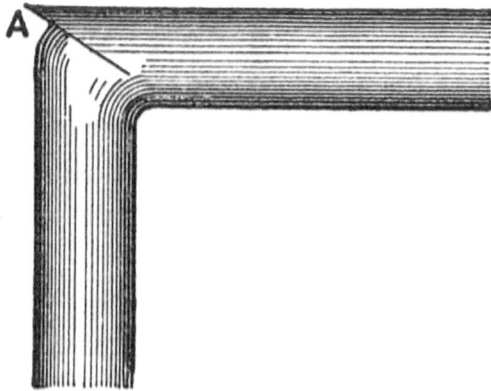

FIG. 110.

FIG. 111.

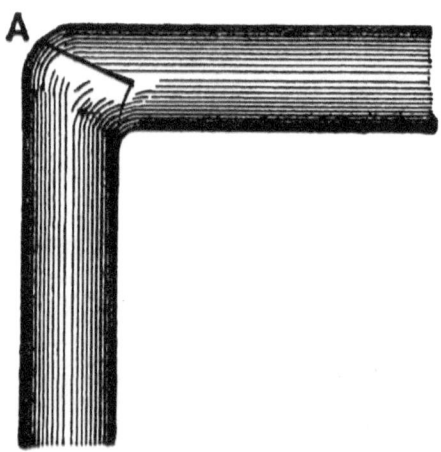

FIG. 112.

WELDING SHAFTS TO AN EXACT LENGTH.

An old blacksmith gives the following method of welding a shaft to an exact length, which he says he has used with unfailing success for many years.

Cut the ends of the two pieces to the exact length the shaft is to be when finished, leaving the ends quite square. Each end is then cut out as in Fig. 113, the length and depth of the piece cut out being equal in all cases to one-half the diameter of the shaft. The shoulder *A*, Fig. 113, is then thrown back with the hammer, and the piece denoted by the dotted line *B* is cut off, leaving the scarf as shown in Fig. 114.

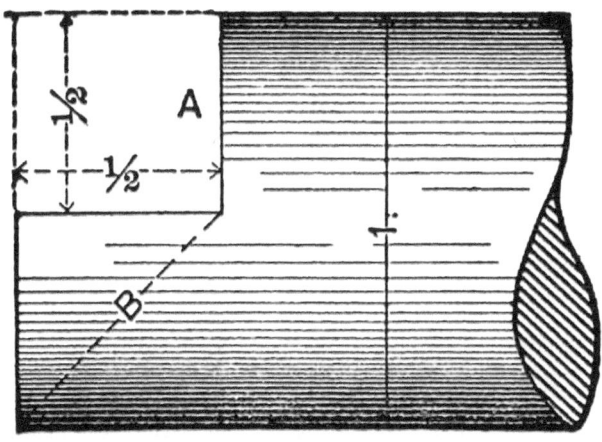

FIG. 113.

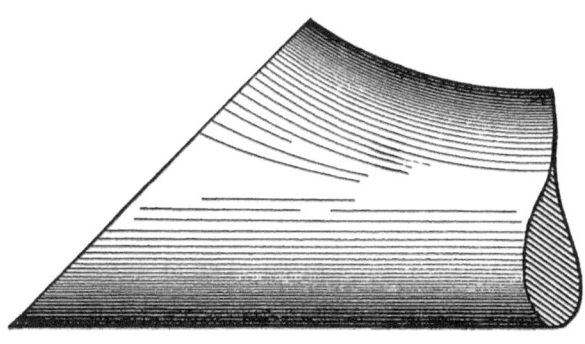

FIG. 114.

The two ends are put together to weld as shown in Fig. 115.

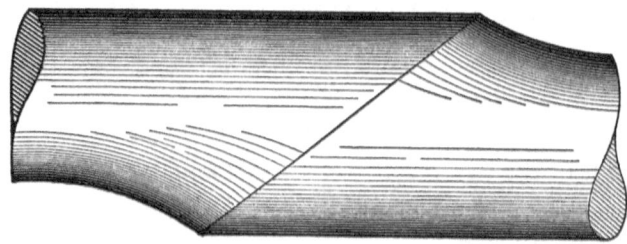

FIG. 115.

The advantage of this plan is that the quantity of metal allowed for wastage during the weld is a known quantity, bearing the necessary proportion to the diameter of the shaft, so that when the weld is swaged down to the proper diameter the length of the shaft will be correct.

WELDING A HEAVY SHAFT.

My plan is, dress the two ends you are going to weld, level, or somewhat rounding, for a four-inch shaft. Then punch a five-eighths-inch hole two inches down each end; then take a piece of five-eighths-inch round iron four inches long and dowel the two together with the same. Next, place the shaft in the fire and revolve it slowly until the welding heat is obtained, and then by means of sledges or a ram, butt it together while it is in the fire.

The other end must be held firm if the shaft is very long. In doing this job I would prefer to use a ram made by taking a piece of shafting and swinging it in a chair. When the butting together has been accomplished, take the shaft out and swage it. The swage should be in front of the forge so that the shaft will lay in it while heating, and then it can be easily moved for swaging.—*By* L. B. H.

WELDING BOILER FLUES.

In welding boiler flues in a shop where you have only such tools as you make yourself, let the flue be held in a vise, then take a diamond point and cut off square, then cut off a piece of the length desired, scarf both the pieces, one inside and the other outside, drive them together and take a borax heat all around, being careful not to burn a hole through. Open the fire a little on top, keep turning the flue, weld the end of the lap down in the fire lightly, then take the flue out of the fire and round it on an old shaft or any round iron, but don't let the shaft or iron be too large.

To test the work, plug up one end, fill the flue with water, heat the end and put in lime or ashes to anneal it. When it is cold anneal the other end. To put it in the boiler you need an expander and beader.—*By* H. R.

MAKING A WELD ON A HEAVY SHAFT

There is only one kind of a weld that will stand for a stem for oil tools, and that is a forked or "bootjack" weld. The accompanying cut, Fig. 116, shows the mode of handling the shaft. It has to be done with the beam *A*, pulley *P*, and chains *C*, revolving around the lower pulley, and the iron shaft or auger stem as seen in or running through the forge or fire. The crotch should be placed so that the blast shall drive the current of heat sidewise on the crotch. Catching any one of the four chains and drawing it down will revolve the shaft without misplacing the crotch in the fire until you have a welding heat.

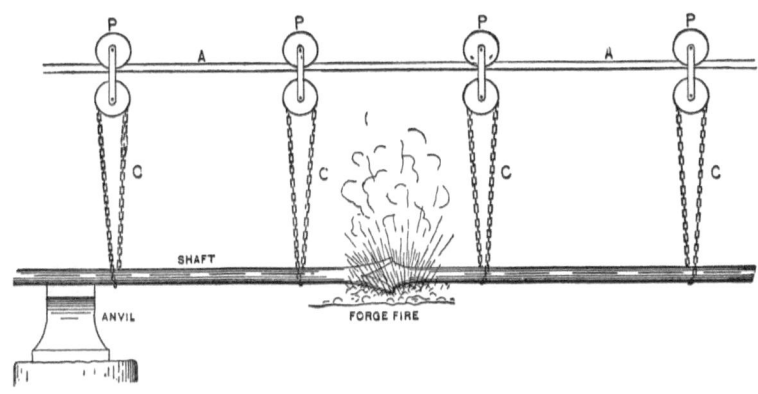

FIG. 116—MAKING A WELD ON A HEAVY SHAFT, AS DONE BY "A. D. G."

Then by drawing the shaft lengthwise, moving the upper pulleys, you can bring the weld to the anvil and with sledge or hammer make your weld.

It has been clearly demonstrated in the oil regions that a weld made with a sledge stands better on heavy stems than one made with a steam hammer. Why this is so I cannot say, but it appears to be a fact that cannot be denied.

I will make a weld after the above plan, and will guarantee that it cannot be jumped, whereas the ordinary lap or buck weld will not stand the continued jars that oil tools have to contend with.—*By* A. D. G.

WELDING ANGLE IRON.

To be a good angle-iron smith is a thing to be proud of, for it requires skillful forging to make good, true, clean work of proper thickness at the weld. Some simple examples are given as follows:

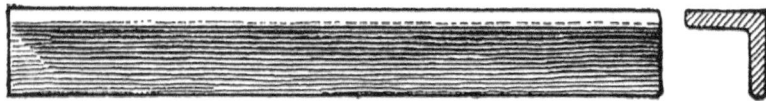

FIG. 117—ANGLE IRON FOR BENDING.

It is required to bend the piece of angle iron shown in Fig. 117 to a right angle.

The first operation is to cut out the frog, leaving the piece as shown in Fig. 118, the width at the mouth *A* of the frog being three-quarter inch to every inch of breadth measured inside the flange as at *B*.

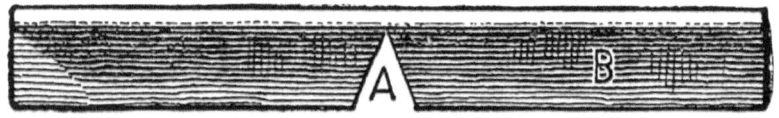

FIG. 118—FIRST OPERATION.

The edges of the frog are then scarfed and the piece bent to an acute angle; but in this operation it is necessary to keep the scarfs quite clean and not to bend them into position to weld until they are ready for the welding heat; otherwise scale will form where the scarfs overlap and the weld will not be sound.

The heat should be confined as closely as possible to the parts to be welded; otherwise the iron will scale and become reduced below its proper thickness.

The iron is then bent to the shape shown in Fig. 119, and the angle to which it is bent is an important consideration.

FIG. 119—BENDING TO SHAPE.

The object is to leave the overlapping scarf thicker than the rest of the metal, and then the stretching which accompanies the welding will bring the two arms or wings to a right angle.

It is obvious, then, that the thickness of the metal at the weld determines the angle to which the arms must be bent before welding.

The thicker the iron the more acute the angle. If the angle be made too acute for the thickness of the iron at the weld there is no alternative but to swage the flange down and thin it enough to bring the arms to a right angle. Hence it is advisable to leave the scarf too thick rather than too thin, because while it is easy to cut away the extra metal, if necessary, it is not so easy to weld a piece in to give more metal. In very thin angle irons, in which the wastage in the heating is greater in proportion to the whole body of metal, the width of the frog at A in Fig. 118 may be less, as say nine-sixteenths inch for every inch of angle iron width measured, as at B in Fig. 118. For angles other than a right angle the process is the same, allowance being made in the scarf-joint and bend before welding for the stretching that will accompany the welding operation.

The welding blows should be light and quick, while during the scarfing the scale should be cleaned off as soon as the heat leaves the fire, so that it will not drive into the metal and prevent proper welding. The outside corner should not receive any blows at its apex; and as it will stretch on the outside and compress on the inside, the forging to bring the corner up square should be done after the welding.

The welding is done on the corner of an angle block, as in Fig. 120, in which *A* is the angle iron and *B* the angle block.

FIG. 120—HOW THE WELDING IS DONE.

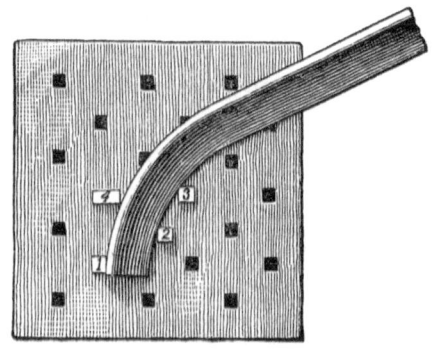

FIG. 121—BENDING ANGLE IRON INTO A CIRCLE.

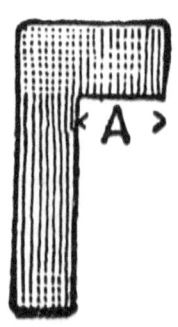   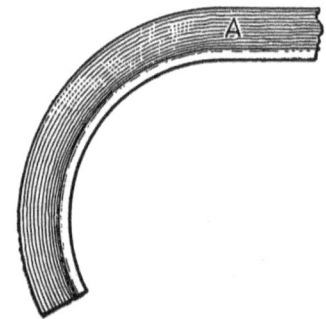

FIG. 122—SHAPE OF PINS.   FIG. 123—SHOWING FLANGE
                             INSIDE OF CIRCLE.

To bend an angle iron into a circle, with the flange at the extreme diameter, the block and pins shown in Fig. 121 are employed. The block is provided with the numerous holes shown for the reception of the pins. The pins marked 1 and 2 are first inserted and the iron bent by placing it between them and placed under strain in the necessary direction. Pins 3 and 4 are then added and the iron again bent, and so on; but when the holes do not fall in the right position, the pins are made as in Fig. 122, the length of the heads A varying in length to suit various curves.

To straighten the iron it is flattened on the surface *A* and swaged on the edge of the flange *B*, the bending and straightening being performed alternately.

When the flange of the angle iron is to be inside the circle, as in Fig. 123, a special iron made thicker on the flange *A* is employed. The bending is accomplished, partly by the pins as before, and partly by forging thinner, and thus stretching the flange *A* while reducing it to its proper thickness.—*By Joshua Rose.*

WELDING COLLARS ON ROUND RODS.

I bend the rod for the collar as shown by Fig. 124, and cut off at *A*. In welding I hit first at B, and go on around to *A*.

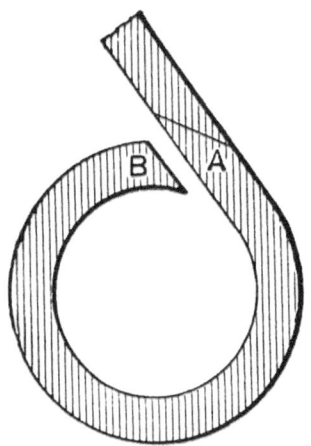

FIG. 124—WELDING COLLARS ON ROUND RODS BY "SOUTHERN BLACKSMITH'S" METHOD. SHOWING HOW THE ROD IS BENT AND CUT OFF.

I weld either at the anvil or on a swage—it doesn't matter which. For large collars I cut off a piece to make the collar, and leave it straight, then I heat the bar and collar separately, and weld one end of the collar first, and bend and weld as I go around, taking separate welding heats after the bending.—*By* Southern Blacksmith.

SHALL SAND BE USED IN WELDING?

The rule among smiths generally seems to be that the more sand they can get on in welding the better, the idea being that the iron will be heated more evenly by this process. Some time ago I got out of sand and could get no more for a time. After I had worked awhile without it I did not want any more, and have not used it since.

Some of the reasons for not using it I will name. The forging is cleaner, and it takes less time to do work without than with sand. I have done just as good work since I stopped using it as before, and have done it in less time.

The sand used by me is from molders' castings.—*By* W. B. G.

FLUXES OR WELDING COMPOUNDS FOR IRON OR STEEL.

No. 1. Here is a welding compound equal to *cherry heat*:

Take ten ounces borax, one ounce muriate of ammonia, pound them roughly together, put them in an iron vessel, and roast them over the fire till all spume has disappeared, then turn it out to cool, afterwards pulverize and keep in a close tin box. Use about like borax, only not so much. I use this all the time and consider it equal to any compound in the market.

I can weld any kind of steel I ever tried with it—springs, fork tines, etc.—*By* J. C. McM.

No. 2. I will give a recipe for welding steel that I have used twenty years successfully: one ounce of copperas, half an ounce of saltpetre, quarter of an ounce of sal ammoniac, three ounces of salt, one and a half pounds of sand. Pulverize and mix together and keep in a dry place.—*By* R. S.

No. 3. An excellent welding compound for steel is composed as follows: two ounces of copperas; four ounces of salt; four pounds of white sand.

Mix the whole and throw it on the heat as is done with sand only.

No. 4. Equal quantities of borax and pulverized glass, well wetted with alcohol and heated to a red heat in a crucible. Pulverize when cool, and sprinkle the compound on the heat.

No. 5. Take copperas, two ounces; saltpetre, one ounce; common salt, six ounces; black oxide of manganese, one ounce; prussiate of potash, one ounce. Pulverize these ingredients and mix with them three pounds of nice welding sand. Use this compound as you would sand. The quantity I have named will cost twenty cents and last a year.—*By* T.

No. 6. Take one part of lime to two or three parts of river sand, such as a plasterer would use for a finishing coat.—*By* R.

No. 7. One part copperas, two parts salt, four parts sand. Thoroughly mix. This makes a splendid welding compound.—*By* A. G. C.

No. 8. Take one ounce of carbonate of iron and mix it with one pound of borax. In using it on plows, always fit the lay very close to the land side and rivet it to get a good weld.—*By* G. W. P.

No. 9. I offer below a recipe for welding steel without borax:

Copperas, two ounces; saltpetre, one ounce; common salt, six ounces; black oxide of manganese, one ounce; prussiate of potash, one ounce. All should be pulverized and mixed with three pounds of nice welding sand. With this preparation welding can be done at a cherry heat.—*By* Anxious.

No. 10. I like this better than borax, but it takes a little more time when it is used. It is cheaper than borax. Those who want to try it can make it by taking a quarter of a pound of rock saltpetre, and dissolving it in a quarter of a pound of oil of vitrol and adding to one gallon of rain water.

After scarfing the steel get it red hot and quench it in this preparation, then heat and weld the same as iron, hammering very quickly with light blows. Keep the compound in stone jars with a tight, fitting lid, and it will be good for years.

SAND IN WELDING—FACING OLD HAMMERS.

No. 11. I wish to say a few words with respect to the use of sand in welding. The question seems to be shall we or shall we not use sand in making welds. I consider it a very essential point in working steel, and use a composition, which I prepare as follows: Take a quart of quartz sand, one pint of common salt, one pint of pulverized charcoal, half a pound of borax well burnt. These I mix well together in a sand box, and consider the preparation much better than raw borax for working steel. In working iron I omit the borax from the compound.—*By* E. T. Bullard.

COMPOSITION FOR WELDING CAST-STEEL.

Borax, ten parts; sal ammoniac, one part; grind or pound them roughly together; fuse them in a metal pot over a clear fire, continuing the heat until the spume has disappeared from the surface. When the liquid is clear, pour the composition out to cool and concrete, and grind to a fine powder, and it is ready for use.

To use this composition, the steel to be welded should be raised to a bright yellow heat; then dip it in the welding powder and again raise it to a like heat as before; it is then ready to be submitted to the hammer.

BRAZING CAST-IRON.

"What is the reason that I cannot braze cast-iron?" asked a machinist the other day. "Every time I try, I fail. Sometimes the cast-iron burns away, and sometimes the brass will not stick. What is the reason?"

Cast-iron may be easily brazed, if, like doing other peculiar jobs, "you know how to do it." Have the iron clean; make it free from grease and acids, which may be injurious; choose any soft brass, or make some for this purpose. The yellow brass used in brazing copper will do; it must contain a large percentage of zinc, or its melting point will be not much lower than that of the cast-iron itself.

Put on the borax before heating the iron. Dissolve the borax, and apply the solution freely to the parts to be brazed. By doing this before heating, a film of oxide is prevented from forming upon the iron. Fasten the parts together, and heat in a clear charcoal fire. Soft coal is not suitable; there is too much sulphur in it.

Heat the work gradually. Apply heat to the largest piece, and keep that piece the hottest. Sprinkle on powdered borax and brass filings, and use plenty of borax. Watch carefully, and get the iron up to a red heat before any of the brass melts. The brass will not adhere unless the iron is hot enough to melt the brass.

Be very careful not to get the iron too hot, or away it melts and the job is lost. When the brass "runs," remove from the fire immediately, and wipe off the superfluous brass, cool off slowly, and finish up the joint.—*American Machinist*.

BRAZING FERRULES.

Chamfer the ends of the piece to be brazed on opposite sides, and file them so that the iron will be bright and clean, bend and let the ends lap about one-eighth of an inch; then lay on a piece of brass. (I use old lamp tops or burners.) Put on a little pulverized borax with a stick or finger, throw on a few drops of water, and it is ready for the fire. Of course all that remains to be done is to melt the brass, and the ferrule is finished. You may or may not, as you like, dip it in water and cool immediately, as it only makes the brass the softer. Hammer on it or drive it, and it will not break in the brazed part. I have made this summer at least two hundred, and have yet to break the first one. They will stand more hammering than will the solid iron.—*By* A. L. D.

BRAZING A FERRULE.

I will describe my way of brazing ferrules.

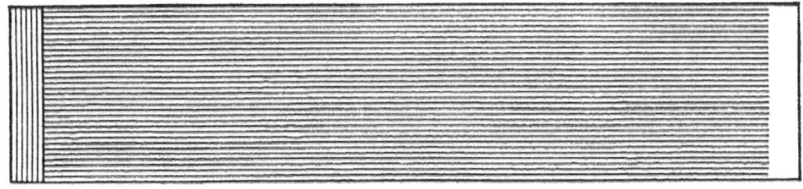

FIG. 125—BRAZING FERRULES BY THE METHOD OF CHAS. W. KOHLER. SHOWING HOW THE EDGES ARE FILED.

If a lap is wanted I file both edges sharp, as in Fig. 125 of the accompanying illustrations, and then make the bend as shown in Fig. 126, put a strip of copper or brass inside, apply burnt borax pulverized, place the ferrule in a clear fire and keep it there until I see a clear flame and then take it out to cool.

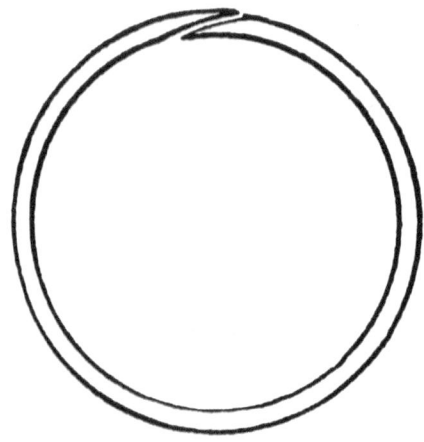

FIG. 126—SHOWING HOW THE FERRULE IS BENT.

If I wish to make the ferrule without a lap, I file the ends square, bend around to the proper shape, bind it with iron wire so it won't open when it gets warm in the fire, and then proceed as in making one with a lap.—*By* C. W. Kohler.

BRAZING.

You cannot braze with cast brass filings. Use granulated "brazier spelter," or sheet brass clippings; white sheet metal for bright work. Wet with water the part to be brazed, then apply powdered borax. The water holds the borax until it calcines or slakes. Then lay on your brazing material. Use a clean coke fire, and as the metal melts poke it with a *wet*, pointed iron where you want it, and remove from the fire at once. Before the flux gets cold, scrape it off with a file or sharp cold-chisel and trim it off. Continue this process until all the flux is removed, heating slightly occasionally if necessary.—*By* Colored Blacksmith.

BRAZING AN IRON TUBE.

If it is anything very particular, I should make two small holes in the tube and two small holes in the piece of iron, and pin them together and wrap binding wire around them. Then take a piece of iron, say one and one-quarter inches thick by two inches wide, and make in it a slot longer than the piece of iron to be brazed. Lay your tube on the iron sideways, and you will see what packing you want under the iron that you are going to braze on to the tube. The larger the tube, the thicker the packing it will require if you want to get it on straight. I think if you follow the directions given you will come out all right. There may be a better way of doing it, but I know of none.—*By* G. P.

BRAZING A BROKEN CRANK

It may be worth while for me to describe my way of brazing. It is as follows: I fasten the broken parts together just as they were before being broken. For example: The accompanying illustration, Fig. 127, represents a corn sheller crank broken at two places in the eye.

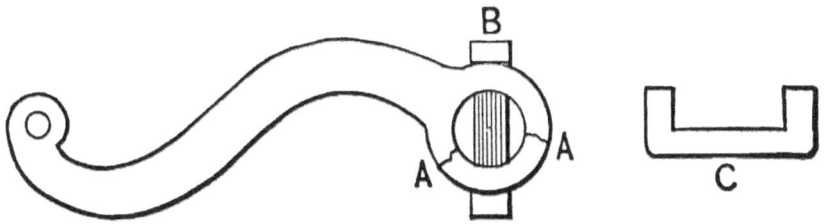

FIG. 127—BRAZING. SHOWING A METHOD OF MENDING A BROKEN CRANK.

A A indicate the places where the breaks exist. I dampen the broken ends, dip in very fine borax and then put on clamp *C*, which is made of iron as thick as the eye of the crank, and is put on the crank hot, the latter being cold. The hot clamp will make the crank swell and this will prevent the clamp from dropping off. I use plenty of brass or brazing solder, or pieces of an old clock frame, but never use cast brass, for it is useless for brazing. I use plenty of borax with the brass, but apply the brass first because if borax is put on first it will swell. When the brass is dropped in the fire and melts, the heat will look like a steel weld. After taking it out of the fire I dip it carefully in the water.—*By* I. N. Bailey.

BRAZING WITH BRASS OR COPPER.

To braze brass or copper, I file the parts to be joined clean, wire them in place or rivet them; then take a few lumps of borax and burn them on a piece of sheet iron, then pulverize them, dissolve some, dip the article to be brazed in it, then lay on the piece of brass or copper, tie it fast, sprinkle some of the borax over it, and put it in a clear fire, blowing very slow at first till the iron gets red; then I will see a blue flame, which is the melting point of brass and copper. I allow it to lie in the fire a minute with blowing, then take it out, and lay it down gently on the hearth to cool.

Very delicate articles should be dipped in a batter of clay to keep them from burning. When the clay begins to glaze it is time to take them out of the fire.

Brass or copper should be brazed with silver. Copper can be brazed with brass, but the melting points of copper and brass are only a few hundred degrees apart, and so such work is not safe unless you have to deal with a large piece of copper. Brass and copper for brazing should be milled. When silver is used it should be old coin or Mexican coin, the silver being purer. The blowpipe is the best to braze with, but it requires some practice to use it successfully.—*By* Chas. W. Kohler.

SOLDERING FLUIDS.

Some of the soldering fluids used are injurious to tools, and also to parts that have been laid on the bench where such fluids have been used. The following recipe will do the work well, and will not rust or tarnish any more than water would: Take two ounces of alcohol and put it into a bottle, and add about a teaspoonful of chloride of zinc and shake until dissolved. Use it in the same manner as the muriate of zinc, or muriatic acid and zinc. It has no bad smell.

A good flux for soldering iron, brass, etc., is made by dissolving chloride of zinc in alcohol.

CHAPTER IV.

STEEL AND ITS USES.

TEMPERING, HARDENING, TESTING.

Extracts from a lecture by Henry Seebohm, of Sheffield, Eng.

I fear that the advantages supposed to be derived from the use of manganese in the manufacture of cast-steel are to a large extent illusory. I have frequently conversed with consumers of steel who knew the trade before the introduction of spiegel iron into Sheffield, and it is remarkable how many of them expressed the opinion that the crucible cast-steel now in use is not so good as it was when they were young. Something may, perhaps, be allowed to the illusions of youth. But, nevertheless, I am convinced there is truth in the opinion that the quality of cast-steel has degenerated. In the present day we sacrifice much to appearances. For my part, I always distrust a bar of steel that has not a "seam" or a "roak" in it. The introduction of manganese into cast-steel is a rough-and-ready way of obtaining soundness at the expense of quality, instead of obtaining it by the tedious care and attention which the steel melter who knows his business gives to each individual crucible.

The question that should come before the consumer of cast-steel is the percentage of carbon which he wishes it to contain. When I first began business the "temper" of steel, or the percentage of carbon which it contained, was concealed from the consumer. The despotic sway of the rule of thumb was absolute. If the consumer discovered that chisel steel contained less carbon than tool steel he owed his discovery entirely to his own wit. My firm was the first to take the consumer into our confidence, and the success which has attended our efforts, and the extent to which our labels have been imitated, have completely justified our act. We have always labeled the steel we supplied to consumers with the percentage of carbon it contained, and the purposes to which, in our opinion, steel containing such percentage of carbon was applicable. The following is a list of the most useful "tempers" of cast-steel:

Razor Temper (one and a half per cent carbon). — This steel is so easily burnt by being overheated that it can only be placed in the hands of a very skillful workman. When properly treated it will do twice the work of ordinary tool steel for turning chilled rolls, etc.

Saw-file Temper (one and three-eighths per cent carbon). —This steel requires careful treatment, and although it will stand more fire than the preceding temper should not be heated above a cherry red.

Tool Temper (one and one-fourth per cent carbon). —The most useful temper for turning tools, drills and planing-machine tools in the hands of ordinary workmen. It is possible to weld cast-steel of this temper, but not without care and skill.

Spindle Temper (one and one-eighth per cent carbon).—A very useful temper for mill-picks, circular cutters, very large turning tools, taps, screwing dies, etc. This temper requires considerable care in welding

Chisel Temper (one per cent carbon).—An extremely useful temper, combining, as it does, great toughness in the unhardened state, with the capacity of hardening at a low heat. It may also be welded without much difficulty. It is, consequently, well adapted for tools, where the unhardened part is required to stand the blow of a hammer without snipping, but where a hard cutting edge is required, such as cold chisels, hot salts, etc.

Set Temper (seven-eighths per cent carbon).—This temper is adapted for tools where the chief punishment is on the unhardened part, such as cold sets, which have to stand the blows of a very heavy hammer.

Die Temper (three-fourths per cent carbon).—The most suitable temper for tools where the surface only is required to be hard, and where the capacity to withstand great pressure is of importance, such as stamping or pressing dies, boiler cups, etc. Both the last two tempers may be easily welded by a mechanic accustomed to weld cast-steel.

We may divide consumers of steel into three classes. First, those who use their own judgment of what percentage of carbon they require, and instruct the manufacturer to send them steel of a specified temper; second, those who leave the selection of the temper to the judgment of the manufacturer, and instruct him to send them steel for a specified purpose; and third, those who simply order steel of a specified size, leaving the manufacturer to guess for what purpose it is required. Fortunately, the size and shape generally furnish some clue to the purpose for which it is likely to be used. For example, oval steel is almost sure to be used for chisels, and small squares for turning tools. One and one-fourth square may be used for a turning tool or a cold set, one and one-fourth round for a drill or a boiler-cup, and the manufacturer has to puzzle his brains to discover whether the chances are in favor of its going into the lathe-room or the blacksmith's shop. It cannot too often be reiterated of how much importance it is, when ordering steel, to state the purpose for which it is going to be used.

When the steel has arrived in the user's hands, the first process which it undergoes is the forging it into the shape required. This process is really two processes. First, that of heating it to make it malleable, and second, that of hammering it, while it is hot, into the required shape. The golden rule in forging is to heat the steel as little as possible before it is forged, and to hammer it as much as possible in the process of forging.

The worst fault that can be committed is to overheat the steel. When steel is heated it becomes coarse grained; its silky texture is lost, and it can only be restored by hammering or sudden cooling. If the temperature be raised above a certain point, the steel becomes what is technically called "burnt", and the amount of hammering which it would require to restore its fine grain would reduce it to a size too small for the required tool, and the steel must be condemned as spoiled. Overheating in the fire is the primary cause of cracking in the water.

The process of hammering or forging the steel into the shape required has hardened the steel to such an extent as to make the cutting impossible or difficult; it must consequently be annealed. This process, like the preceding one, is a double process. The steel must be reheated as carefully as before, and afterward cooled as slowly as possible.

We now come to the culminating point in our manufacture, where the invaluable property which distinguishes steel from wrought iron or cast metal is revealed.

The part of the tool required to be hardened must be heated through, and heated evenly, but must on no account be overheated. Our tool must be finished at one blow—the blow caused by the sudden contraction of the steel produced by its sudden cooling in the water—and if this blow is not sufficient to give to the steel a fine grain and silky texture—if, after the blow is given, the fracture, were it broken in the hardened part, should show a coarse grain and dull color, instead of a fine grain and glossy luster, our tool is spoiled, and must be consigned to the limbo of "wasters." The special dangers to be avoided in hardening each kind of tool must be learned by experience. Some tools will warp or "skeller," as we say in Yorkshire, if they are not plunged into the water in a certain way. Tools of one shape must cut the water like a knife; those of another shape must stab it like a dagger. Some tools must be hardened in a saturated solution of salt, the older the better, while others are best hardened under a stream of running water.

In some tools, where the shape necessitates a great difference in the rapidity of cooling, it is wise to drill holes in the thicker parts where they will not interfere with the use of the tool—holes which are made neither for use nor ornament, but solely with a view of equalizing the rapidity of the various parts, so as to distribute the area of tension and thus lessen the risk of cracking in hardening. So many causes may produce water-cracks that it is often difficult to point out the precise cause in any given case. Perhaps the most common cause is overheating the steel in one or more of the processes which it passes through in the consumers' hands, or it may have been overheated in the process of forging, or rolling it into the dimensions required while in the hands of the manufacturer. A second cause may be found in the over-melting, or too long boiling of the steel, causing it to part with too much of its confined carbonic acid, a fault which may be attributed to the anxiety of the manufacturer to escape honeycomb in the ingot. A third cause may be sometimes discovered in the addition of too much manganese, added with the same motive. A fourth cause may, curiously enough, prove to be a deficiency of carbon, while, in some cases, too much will produce the same effect. A fifth cause may be one which, as a steel manufacturer, I ought to mention in a whisper— the presence of too much phosphorus in the steel; but, after all, this may not be the fault of a greedy manufacturer, who wants to make too great a percentage of profit. It might be the fault of a stingy consumer, who begrudges him the little profit he makes. You may depend upon it there is nothing so dear as cheap steel. It must be more economical to put five shillings' worth of labor upon steel that costs a shilling, to produce a tool that lasts a day, than to put the same value of labor upon a steel that costs only ninepence, to produce a tool that only lasts half a day.

Our difficulties are not quite over when the process of hardening has been successfully accomplished. Our steel was originally lead; it has now become glass. To attain its proper condition our tool must pass through the final process—that of tempering.

If you heat a piece of hardened steel slightly, and allow it to cool again, it becomes tempered. It suddenly changes from glass to whalebone; and in the process of changing its nature, it fortunately changes its color, so that the workman can judge by the hue of the color the extent of the elasticity which it has acquired, and can give to each tool the particular degree of temper which is most adapted to its special purpose. The various colors through which tempered steel successfully passes are as follows: Straw, gold, chocolate, purple, violet and blue. Of course, in passing from one color to another, the steel passes through the intermediate colors. It really passes through an infinite series of colors, of which the six above mentioned are arbitrarily selected as convenient stages.

It is supposed that the maximum of hardness and elasticity combined is obtained by tempering down to a straw color. In tempering steel regard must be had to the quality most essential in the special tool to be tempered; for example, a turning tool is required to be very hard, and is generally taken hot enough out of the water to temper itself down to a degree so slight that no perceptible color is apparent, while a spring is required to be very elastic, and may be tempered down to a blue. If you ask me to give you a scientific explanation of the process of tempering steel, I must confess my absolute ignorance.

Hardening in oil is another mode of treating steel, which appears to a certain extent to attain by one process the change from lead into whalebone without passing through the intermediate glass stage, and is of great value for certain tools.

There are many kinds of steel to which your attention should be called, but which can only obtain from me the briefest mention. A special steel for taps, called mild-centered cast-steel, is made by converting a cogged ingot of very mild cast-steel, so that the additional carbon only penetrates a short distance. These bars are afterward hammered or rolled down to the size required, and have the advantage of possessing a hard surface without losing the toughness of the mild center.

Another special steel, somewhat analogous to mild-centered cast-steel, is produced by melting a hard steel on to a slab of iron, or very mild steel heated hot enough to weld with the molten steel, so that a bar may be produced, one-half of which is iron and the other half steel, or three-fourths iron and one-fourth steel, as may be required.

A third kind of special steel, which is used for turning tools for chilled rolls, magnets and some other purposes, is made by adding a certain percentage of wolfram, or, as the metal is more generally called, tungsten, sometimes with and sometimes without carbon, sometimes to such an extent that it can be used without hardening in water. Special steel of this kind is the finest-grained that can be produced, but it is so brittle that in the hands of any but exceptionally skilled workmen it is useless. The addition of chromium, instead of wolfram, has somewhat the same effect.

TESTING.

It is much to be regretted that no easy method of testing cast-steel has been invented. The amount of breaking strain and the extent of contraction of the area of the fracture are all very well for steel which is not hardened, and not required to be used in a hardened state, but for hardened and tempered steel it is practically useless. It is very difficult to harden and temper two pieces of steel to exactly the same degree. A single test is of comparatively small value, as a second-rate quality of steel may stand very well the first time of hardening, but deteriorates much more rapidly every time it is rehardened than is the case with high quality steel. Nor am I at all sure that the breaking strain is a fair test of the quality of steel. For many tools the capacity to withstand a high amount of breaking strain slowly applied is not so much required as its capacity to withstand a sudden shock. The appearance of the fracture is very illusory. The fineness of the grain and the silkiness of the gloss is very captivating to the eye, but it can be produced by hammering cold. The consumer of steel may be enraptured by the superb fracture of a bar of steel; but, after all, this is only a dodge, depending upon the inclination of the axis of the revolving hammer to the plane of the anvil. The practical consumer of steel must descend from the heights of art and science and take refuge in the commonplace of the rule of thumb, and buy the steel which his workmen tell him is full of "nature" and "body."

HINTS REGARDING WORKING STEEL.

Salt water is no more likely to crack steel than fresh or soft, if the steel has been properly, *uniformly*, heated. Brine will produce a greater degree of hardness at the proper heat, other things being equal.

Steel requires the same conditions for annealing, whether in bulk, ponderous or otherwise. The *sine qua non* is, heat uniformly to proper heat and as *soon* as homogeneously saturated (not super-saturated) permit to cool as *slowly* and uniformly as practicable.

My experience says: Harden in every case where practicable under a stream of cold water, taking care that the contact is perfect and directed to the *locale* requiring the greatest degree of hardening. It is almost always prudent to move the article constantly, if only slightly.

Wherever water is in impinging motion it is, of course, more rapidly changing its heated for a colder co-efficient, or successive heat extractor, we will call it, and the more extractors receiving heat the quicker the locality is refrigerated and fixed, consequently the hardest.

Very few, if any, drop hammer dies, if properly hardened, require a subsequent tempering.

I have known good gray, or, in some cases, white cast-iron, capable of doing twenty times the amount of work of any kind of steel tried.

My way is to heat all *ordinary* brands of "tool cast-steel" very slowly (and if practical) in a cool fire, to commence with, gradually letting (by draught rather than blast) both fire and steel increase in temperature together, to as low red as is necessary.—*By* W. Dick.

THE WARPING OF STEEL DURING THE HARDENING PROCESS.

In heating steel to harden it, especial care is necessary, particularly when the tool is one finished to size, if its form is slight or irregular, or if it is a very long one, because unless the conditions both of heating and cooling be such that the temperature is raised and lowered uniformly throughout the mass, a change of form known as *warping* will ensue. If one part gets hotter than another it expands more, and the form of the steel undergoes the change necessary to accommodate this local expansion, and this alteration of shape becomes permanent. In work finished and fitted this is of very great consideration, and, in the case of tools, it often assumes sufficient importance to entirely destroy their value. If, then, an article has a thin side, it requires to be so manipulated in the fire that such side shall not become heated in advance of the rest of the body of the metal, or it will become locally distorted or warped. If, however, the article is of equal sectional area all over, it is necessary to so turn it in the fire as to heat it uniformly all over; and in either case care should be taken not to heat the steel too quickly, unless, indeed, it is desirable to leave the middle somewhat softer than the outside, so as to have the outside fully hardened and the inside somewhat soft, which will leave the steel stronger than if hardened equally all through. Sometimes the outside of an article is heated more than the inside, so as to modify the tendency to crack from the contraction during the quenching; for to whatever degree the article expands during the heating, it must contract during the cooling. Hence, if the article is heated in a fluid, it may often be necessary to hold the article, for a time, with the thick part only in the heating material; but in this case it should not be held quite still, but raised and lowered gradually and continuously, to insure even heating.

Pieces, such as long taps, are very apt to warp both in the fire and in the water. In heating, they should rest upon an even bed of coked coal, and be revolved almost continuously while moved endways in the fire; or when the length is excessive, they may be rested in a heated tube so that they may not bend of their own weight. So, likewise, spirals may be heated upon cylindrical pieces of iron or tubes to prevent their own weight from bending or disarranging the coils.

If a piece is to be hardened all over, it must be occasionally turned end for end, and the end of the holding tongs should be heated to redness or they will abstract the heat from the steel they envelop. Very small pieces may be held by a piece of iron wire or heated in a short piece of tube, the latter being an excellent plan for obtaining a uniformity of heat, but in any event the heating must be uniform to avoid warping in the fire, and, in some cases, cracking also. This latter occurs when the heating takes place very quickly and the thin parts are not sufficiently heated to give way to accommodate the expansion of the thick ones. The splitting or cracking of steel during the cooling process in hardening is termed *water* cracking, and is to be avoided only by conforming the conditions of cooling to the size and shape of the article.

Experiments have demonstrated that the greater part of the hardness of steel depends upon the quickness with which its temperature is reduced from about 500° to a few degrees below 500°, and metal heated to 500° must be surrounded by a temperature which renders the existence of water under atmospheric pressure impossible; hence, so long as this temperature exists the steel cannot be in contact with the water, or, in other words, the heat from the steel vaporizes the immediately surrounding water. As the heated steel enters the water the underneath side is constantly meeting water at its normal temperature, while the upper side is surrounded by water that the steel has passed by, and, to a certain extent, raised the temperature of. Hence, the vapor on the underneath side is the thinnest, because it is attacked with colder water and with greater force, because of the motion of the steel in dipping. Suppose, now, we were to plunge a piece of heated steel into water, and then slowly move it laterally, the side of it which meets the water would become the hardest, and would be apt to become concave in its length.

From these considerations we may perceive how important a matter the dipping is, especially when it is remembered that the expansion which accompanies the heating is a slow process compared to the contraction which accompanies the cooling (although their amounts are, of course, precisely equal), and that while unequal expansion usually only warps the article, unequal contraction will in a great many, or, indeed, in most cases, cause it to crack or split.—*By* Joshua Rose, M. E.

TEMPERING STEEL.

I have read with considerable interest the different processes for tempering steel, and the different treatments that have appeared from time to time in mechanical journals. It always seemed to me the object to be sought for is to cool the steel as soon as possible after it has been heated to the well known cherry red, and, in doing so, the different solutions offered for a cooling bath are as numerous as the recipes for a toothache. Mercury, no doubt, makes the best bath, and salt water, for many purposes, is equally as good. But one object I have observed in quenching steel, is the time required for reducing the heat. For some moments the heated steel shows its cherry color through the cooling bath. Mercury, being the best conductor of heat, takes away the heat the fastest and hardens it the hardest. If there were any way in which the cold water could be brought in contact with the metal, its heat would the sooner be removed. Warm water hardens almost as well as cold, except for light work. This may be owing to the jacket formed about the heated steel, which protects the steel from losing its heat. This is more particularly noticed in hardening the face of a hammer so that it will not settle in the center, or cave off around its edge. In plunging it into the bath, the center of the face is, to a certain extent, protected by the heat from the outer edge, and remains the softest. The drawing process or tempering being governed by the color of the oxidization that appears on the polished surface, there is no way to distinguish the hard places from the soft that were produced in hardening; and without further comment, will state that the process of hardening by cooling with water that is brought in close contact with the heated surface by pressure, promises well, and should find its way more effectually into general use. The writer remembers very well when at work in a shop close by a reservoir, where a stream of water rushing through an orifice, under a head of twelve feet or more, was always ready for quenching steel for the purpose of hardening. Hammers were heated as near the right temperature as was thought best and held under the stream where the water would strike square upon the face of the hammer, removing the heat with great rapidity, the heated liquid passing off out of the way and the cooler taking its place, and, owing to the great pressure, the film of vapor that might otherwise be formed between the heating and cooling surfaces is broken up. Smooth metallic surfaces, when heated to a low red heat, are protected from coming in contact with a liquid by the intervening film of steam which gives an imperfect in hardening, and is known as the spheroidal state of liquids, as observed when cooling sheets of iron by pouring water upon them, the liquid will run about in large drops without breaking up or boiling. Steel is hardened to a remarkable degree by being forced in contact with almost any cold substance. A flat drill that has been heated at the point, and driven into a lump of cold lead, is as hard as if quenched in a bath of cold water, Any way to remove the heat is all that is required for hardening, and the sooner it is removed the better, and in chilling with cold water it is necessary that the steel should be moved about to break up the film, and to keep in contact with the cooling liquid. I have felt an interest in the matter of tempering steel, and cannot but feel that this state of things in regard to the close communion between heated metals and their cooling solutions should be more fully understood.—*Cotton, Wool and Iron.*

ANOTHER METHOD OF TEMPERING STEEL.

It is desirable to obtain any degree of hardness by a single process if possible. In some cases, by heating a known quantity of steel to a definite temperature and quenching it in liquid maintained at about an even temperature, the color is becoming dispensed with, the conditions of heating and cooling being varied to give any degree of hardness. Another and a very desirable method of hardening and tempering, is to heat in a flue of some kind, maintained at the required temperature over the fire, and after quenching, instead of applying the color test, provide a tempering bath composed of some substance heated to a temperature of from 430° to 630°. By placing the articles (after hardening them) in the tempering bath and heating it to a temperature equal to the color of the temper required, we have but to cease heating the tempering bath when a thermometer marks the required temperature. A uniform degree of temper will be given to all the articles, and the operation will occupy much less time than would tempering by the color test, because a liquid is much more easily kept at an equal temperature throughout its mass than are the heated sand or hot pieces of metal resorted to in tempering by the color test. Another method of tempering is to heat the steel to a definite temperature and cool or quench it in a liquid having sufficient greasiness or other quality which acts to retard its retraction of the heat from the steel and thus give a temper at one operation. As an example of this kind of tempering, it is said that milk and water mixed in proportions determined by experiment upon the steel for which it was employed, has been found to give an excellent spring temper. Such tempering carefully conducted may be of the very best quality. A great deal, however, in this case depends on the judgment of the operator, because very little variation in heating the steel or in the proportions of milk to water produces a wide variation in the degree of temper. If, on trial, the temper is too soft, the steel may be made hotter or more water added to the milk. If the steel was heated as hot as practicable without increasing the danger of burning it, more water must be added, while if the steel was made red-hot without being hot enough to cause the formation of clearly perceptible scale, the steel may be heated more. It is desirable in all cases, but especially with a high quality of steel, not to heat it above a blood-red heat, although sheer and spring steels may be and often must be made hotter in order to cause them to harden when quenched in water.

Hardening and tempering steel, as applied to cutting tools, are much more simple than when the same operations are required to give steel elasticity as well as durability of form or to give durability to pieces of slight and irregular form of sufficient hardness to withstand abrasion. One reason of this is that for tools a special and uniform quality of steel is readily obtainable, which is known as tool steel. Special sizes and grades are made to suit the manufacture of any of the ordinary forms of tools.

As a rule, the steel that shows a fracture of fine, dull grain, the face of the fracture being comparatively level, is of better quality than that showing a coarse or granulated surface, brightness denoting hardness, and fibrousness, toughness.

Very few steels are as yet sufficiently uniform to render it practicable to employ an unchangeable method of tempering, and to this fact is largely due the use of particular brands of steel.

In tempering steel, regard must be had to the quality most essential in the special tool to be tempered. A turning tool is required to be very hard and is often taken out of the water hot enough to temper itself down to a degree so slight that no color is perceptible, while a spring is required to be very elastic and may be tempered down to a blue.

A scientific explanation of the process of tempering steel has yet to be given without mystifying one by talking unintelligibly about molecular rearrangement and crystalline transportations.

WORKING STEEL.

In making steel tires blacksmiths generally heat them too much. A deep cherry-red is hot enough. Of course you can't make much headway in hammering, but you should heat oftener. It is better to spend time in heating often than to burn the steel, spoil the job and get nothing for it. When steel has been hammered cold and gets black it should never be heated hot enough to raise a scale, because this would open the pores and render it worthless for anything requiring tempering. Heat just to a low cherry-red and draw your temper accordingly. Some smith may say: "It wouldn't be hard enough." Don't be afraid of that; it will be hard enough for anything.—*By* Benninger & Son.

WORKING AND TEMPERING STEEL.

To work steel never heat above a light cherry-red for hammering, then hammer light and quick until near black, as this improves the steel and will make tools that will do more than double the work than if not so treated. The hardness of steel is governed entirely by the heat when it is dipped in water; for instance, a piece of steel dipped at a bright cherry color and drawn to a straw, will be very much harder than a piece heated to a dark cherry-red and then dipped and drawn to a straw. Try it.

The forging, hardening and tempering of steel is an art that but few understand, as its knowledge is only gained by experience, and but few ever give its secrets to others; yet in a few words I will try to give the principal elements to workers of steel, which if followed will save you many losses, and give you a reputation for working steel that will ensure you good and serviceable tools, as well as increase your gains.

Please remember that the heat at which steel is worked and hardened are two of the vital elements to produce good and serviceable tools. If heated above a light cherry-red, some of the vitality of the steel is - destroyed, and it would in heating too many times return to iron. If heated too hot when hardening it would fly to pieces, destroying your labor and steel as well as giving you a poor reputation.

Remember also to hammer your work lightly at a low heat, as this improves chisels, drills, lathe tools, and edge tools most wonderfully; also take as few heats as possible, as overheating and too frequent heating reduces the steel to iron by destroying the carbon.

To harden taps, rimmers, chisels and drills, always dip them slowly to the depth desired in as near a vertical line as you can by the eye and hand, then move in a circular position until cold, but never any deeper in the water than first dipped, as this prevents them from cracking, which they would be likely to do if held perfectly still and the water formed a line around them. Do not change the water in which you temper, but as it wastes fill up the tank. If you are obliged to use fresh water always heat a piece of iron to put into it and bring it to such a warmth as is perceptible to the hand, as steel is liable to crack when dipped into cold water, When you have heated your article to be tempered take it from the fire and examine to see if any flaws are observable in the steel, as this will prevent your having poor pieces of steel laid to your carelessness in hardening.

In cutting up steel a thin, sharp chisel should be used, as a blunt one is liable to splinter or crack the bar, which will not be seen until it is tempered and then the labor is lost with the steel.

Colors of different articles for use. —Taps should be hardened and then brightened by rubbing emery and oil on the clearance, and then draw on a hot plate or in a heated ring to a dark straw color.

Dies should be a bright straw color and drawn on a hot plate or in sand.

Drills for iron should be a dark straw on the cutting part and the rest a blue.

Chisels for iron should be violet color; for cutting stone a purple is required.

Milling cutters should be of a yellowish white. Gear teeth cutters the same color. The usual way to dip these is to have a rod with three prongs to pass through the hole after it is heated to dip with, lower slowly until all the cutter is under the water about two inches, then move in a circular position until thoroughly cold, remembering that a great many things break by taking from the water before they are cold, especially large pieces of steel, as the center retains the heat, and when taken from the water it expands the outside and causes it to crack.

In tempering pieces having a thick and thin edge, always dip the thickest part first. Study the pieces you have to harden and it will help you very much. Large centers in work for tempering should be avoided, as they are liable to cause the end to split open.—*Iron Trade Review*.

TEMPERING STEEL.

Two of the most important processes in blacksmithing, are the hardening and the tempering of steel. Great judgment particularly, as well as experience, is required to temper dies and tools. With good judgment, a person will soon learn to temper; but without good judgment, tempering can never be successfully learned. A man may learn to do one special kind of work, but put him in a large hardware manufacturing establishment, where all kinds of dies and tools are used, with hardly any two requiring the same temper, and without judgment, the difficulties connected with such a position can never be overcome.

To harden and temper a piece of steel, it should always be properly annealed; otherwise it is almost certain to spring or warp. It is a very general idea that you can draw all the temper out of steel, without heating it red hot all over; but such is not the case. Heating the face of a die and covering it up in ashes does not thoroughly anneal it by any means. Possibly in that way you can get it in such a condition that it can be worked; but it will not be very soft, and will not harden and temper as well as if it had been heated all over.

The best way that I ever found to anneal steel (when you do not have a kiln for that purpose) is to heat it all over in a slow charcoal fire; the slower it is heated the better. Do not heat very hot, but all over and all through. Then take it out and cover it up in fine charcoal, and let it remain till cold.

In heating a die for the purpose of hardening, it is not necessary to heat it all over, unless you want to harden it all over. The only way, in my judgment, that dies can be tempered as they should be, is for the one that tempers them to see, from time to time, how they work, as they are being used; and in that way he can tell if they require more or less temper and the particular places where they should be hard or soft.

It is useless for anyone to undertake to tell how to temper everything that is used in a manufacturing establishment; such knowledge can only be acquired by experience, combined with good judgment and mechanical ingenuity. I will try, however, to explain my way of hardening and tempering.

Trip-hammer dies may seem to be very easy to handle, but a blacksmith without experience would meet with many difficulties. Take, for instance, a die seven or eight inches long, one and a half inches thick and four and a half inches wide with several impressions cut in it, leaving several small points which are liable to fly off as soon as they are hardened or when you are drawing the temper. Now, my way to prevent the corners from coming off, and keep the dies straight is this: I heat my die very carefully in a clean charcoal fire, being careful not to get it any hotter than is necessary to have it harden. When the proper heat has been obtained, grasp it with the tongs near one end in such a way that it can be put in the water perpendicularly, and while in the water I turn it to a horizontal position and take it about half-way out, letting it remain until it is cool enough on the face to take out entirely for an instant without permitting the temper to run down. Then I withdraw it and return it to the water several times very quickly until it is cool enough to take out and scour off. The temper can then be drawn to suit, without any danger of the corners flying off.

The object in putting it in the water perpendicularly is to keep it straight; and, as it cools off all alike, when you bring the back out of the water the heat rushes up toward the back, and expands it, taking all the strain off the face of the die and preventing it from breaking.

By plunging it into the water and withdrawing it very quickly, the face of the die has time to get cool gradually, and the corners are thus prevented from flying off.

This process will work well on any dies or tools hardened in this way. I have tried it hundreds of times with the best of success.—*By* G. B. J.

TEMPERING SMALL ARTICLES.

When tempering cold chisels, or any other steel articles, heat to a very dull red and rub with a piece of hard soap, then finish heating and harden in clear, cool water. The potash of the soap prevents the oxygen of the atmosphere from uniting with the steel and forming rust or black oxide of iron. The article will need no polishing to enable the colors to be seen. This will be appreciated when tempering taps, dies or various complex forms not easy to polish. Never "upset" a cold chisel. It is sure death to steel. Many of us have lived on a farm and know something about a bundle of nice, straight, clean straw. If you work it intelligently you can tie it up into stout bands for binding other bundles. You can take hold of the ends of the straw and draw out a handful without harm to the straw. After you have drawn out half that bundle a foot or so, try to drive it back; every blow breaks the straw, cripples and doubles it up, and it will hardly bear its own weight, to say nothing of making a band for other bundles. Just so with steel. If you have a broken chisel to sharpen, draw out and cut off, never upset. It will cripple the fibers just as the straw is crippled when driven endwise.

Make chisels short for hard, rough work. They transmit the power or force of a blow much better. Long chisels are apt to "broom up" on the hammer end, as the long steel through which the blow passes has more chance to absorb the force of the blow.

The harder the metal to be worked, the quicker the blow should be transmitted. Cast-iron works much better with a short steel chisel and light hammer, than if the blow was struck upon a very long chisel with a heavy wooden mallet.—*Age of Steel.*

TEMPERING STEEL.

I have never used any mercury in tempering but have no doubt it answers the purpose. I was one of the first blacksmiths in this country who worked cast-steel. The boss under whom I learned the trade made it a part of his business to teach country blacksmiths how to put cast-steel in axes and how to weld and temper it. He welded with borax melted into a kind of hard glassy substance.—*By* C. W.

TEMPERING STEEL WITH LOW HEAT.

Some curious statements on tempering steel are made in a paper published in *Dingler's Polytechnic Journal*, vol. 225, by Herr A. Jarolimek, "On the Influence of the Annealing Temperature upon the Strength and the Constitution of Steel." Hitherto it has been generally considered that to obtain a specified degree of softness it is necessary to heat the hard steel to a particular annealing color—that is to say, to a definite temperature—and then allow it to rapidly cool. Thus, for example, that steel might anneal—be tempered—yellow, it has to be heated to 540 deg. and the supposition was formed and acted upon that it must be allowed only a momentarily subjection to this temperature. Herr Jarolimek says the requisite temper, which is obtained by momentarily raising the temperature to a particular degree, can also be acquired by subjecting the steel for a longer time to a much lower temperature. For example, the temper which the annealing color—yellow—indicates, can be obtained by exposing the hard steel for ten hours to 260 degrees of heat; in other words, by placing it in water rather above the boiling point.

TO TEMPER STEEL VERY HARD.

As hardness of steel depends on the quickness with which it is cooled, there are better materials than water, which gives an unequal temper; besides the steam bubbles developed interrupting contact; water is also a bad conductor of heat, and if the bubbling and heat did not put it in motion, it would be unfit for hardening. Water with plenty of ice in it gives a hard temper; small tools may be stuck into a piece of ice, as jewelers insert them in a piece of sealing wax. Oil is also used by them as being better than water, as it does not evaporate so easily. The Damascus steel blades are tempered in a small current of cold air passing through a narrow slit; this gives a much more uniform and equal temperature than water. But the most effective liquid is the only liquid metal—mercury. This being a good conductor of heat, in fact the very best liquid conductor, and the only cold one, appears to be the best one for hardening steel-cutting tools. The best steel, when forged into shape and hardened in mercury, will cut almost anything. We have seen articles made from ordinary steel, which have been hardened and tempered to a deep straw color, turned with comparative ease with cutting tools from good tool steel hardened in mercury. Beware of inhaling the vapor while hardening.

HARDENING STEEL.

Often when great care has been taken in heating a straight piece of steel, and it is put into water or other hardening compound, it comes out crooked; in this case the trouble is entirely in the forging. I will give my reasons for this statement.

Long pieces of steel, such as reamers for boring boxes for axles, are generally forged under a trip hammer. We will suppose the piece of steel to be forged is evenly heated through. The blacksmith takes the bar in one hand, and in the other his hammer, and the helper holds his sledge ready for business. The smith turns the bar back and forth, never turning it entirely over. Now, the hammer and the sledge will draw out the fiber or grain of the steel faster than the anvil. The steel is unevenly forged, and very likely will not be straight, but will be made straight across the anvil. Heat this piece of steel as evenly as possible, and put it into water or other compound, and it will very likely be crooked when hardened.

If a piece of steel is heated evenly, and hammered equally as many blows on all sides, and if, when crooked, while forging, it is straightened by hammering, and care has been used in heating, it will generally come out straight, providing the same care has been observed before it comes to the blacksmith.—*By* C.

CASE-HARDENING STEEL OR IRON.

I think the best and simplest method of case-hardening iron is to use prussiate of potash. It is a yellow substance that comes in the shape of flaky, shining flat lumps. Pulverize it till it is as fine as flour; heat the article to be hardened to a deep red, and put on the potash just where you want it to be hard. Then put it back in the fire, heat to a deep red and plunge into cold water.—*By* J. P. B.

HOW DAMASCUS SWORD BLADES WERE TEMPERED.

Perhaps the best method which has ever been discovered for tempering steel, resulting in hardness, toughness and elasticity combined, is that followed in hardening the blades of the famous Damascus swords. The furnace in which the blades were heated was constructed with a horizontal slit by which a current of cold air from the outside entered. This slit was always placed on the north side of the furnace and was provided on the outside with a flat funnel-shaped attachment by which the wind was concentrated and conducted into the slit. The operation of tempering the blades was only performed on those days of winter when a cold strong north wind prevailed. The sword blade when bright red-hot was lifted out of the fire and kept in front of the slit and by this means was gradually cooled in the draft of air. It acquired the proper degree of temper at the single operation.

TO HARDEN STEEL.

A very fine preparation for making steel very hard is composed of wheat flour, salt and water, using say two teaspoonsful of water, one-half a teaspoonful of flour, and one of salt; heat the steel to be hardened enough to coat it with the paste—by immersing it in the composition—after which heat it to a cherry-red and plunge it in cold, soft water. If properly done, the steel will come out with a beautiful white surface. It is said that Stubb's files are hardened in this manner.

TEMPERING STEEL SPRINGS.

The hardening and tempering of springs whose coils are of thick cross-section is performed at one operation as follows: The springs are heated in the furnace or oven described, and are first immersed for a certain period in a tank containing fish oil (obtained from the fish *"Moss Bunker"*, and termed *"straights"*), and are then removed and cooled in a tank of water. The period of immersion in the oil is governed solely by the operators judgment, depending upon the thickness of the cross-section of the spring coil, or, in other words, the diameter of the round steel of which the spring is made. The following, however, are examples:

Number of coils in spring	5 ¾
Length of the spring	6 inches.
Outside diameter of coils	4 ¾ "
Diameter of steel	1 inch.

The spring was immersed in the oil and slowly swung back and forth for twenty-eight seconds, having been given thirty-five swings during that time. Upon removal from the oil the spring took fire, was re-dipped for one second, and then put in the cold water tank to cool off.

Of the same springs the following also are examples:

Example	Time of immersion in oil	Number of swings in oil.
Second	36 seconds	46
Third	27 " "	36
Fourth	38 " "	40

SIZE OF SPRING.

Number of coils in the springs	6
Length of the springs	9 inches.
Inside diameter of coils	3 ¾
Size of steel	1 x 1 ½ square.

Example	Time of immersion in oil	Number of swings in oil.
First	9 seconds	12
Second	8 "	12
Third	8 "	12
Fourth	9 "	12
Fifth	9 "	12
Sixth	9 "	12

To keep the tempering oil cool and at an even temperature, the tank of fish-oil was in a second or outer tank containing water, a circulation of the latter being maintained by a pump. The swinging of the coils causes a circulation of the oil, while at the same time it hastens the cooling of the spring. The water-tank was kept cool by a constant stream and overflow.

If a spring, upon being taken from the oil, took fire, it was again immersed as in the first example.

In this, as in all other similar processes, resin and pitch are sometimes added to the oil to increase its hardening capacity if necessary.

The test to which these springs were subjected was to compress them until the coils touched each other, measuring the height of the spring after each test, and continuing the operation until at two consecutive tests the spring came back to its height before the two respective compressions. The amount of set under these conditions is found to vary from three-eighths of an inch, in comparatively weak, to seven-eighths of an inch for large, stiff ones.

The springs were subjected to a severe test in a machine designed for that purpose, being compressed and released until there was no set under the severest test.

In the following description of the plans adopted by a very prominent carriage-spring maker will also be found a process termed a water-chill temper, which tempers at one process. The steel used by this firm is "Greave's spring steel." The spring plates are heated to bend them to the *former*, which is a plate serving as a gauge whereby to bend the plate to its proper curve.

This bending operation is performed quickly enough to leave the steel sufficiently hot for the hardening; hence the plates after bending are dipped edgeways and level into a tank of linseed oil which sets in a tank of circulating water, the latter serving to keep the oil at about a temperature of 70 degrees when in constant use. About three inches from the bottom of the oil tank is a screw to prevent the plates from falling to the bottom among the refuse sediment.

To draw the temper the hardened springs are placed in the furnace, which has the air-blast turned off, and when the scale begins to rise, showing that the adhering oil is about to take fire, they are turned end for end in the furnace so as to heat them equally all over. When the oil blazes and is freely blazed off, the springs are removed and allowed to cool in the open air, but if the heat of a plate, when dipped in the oil to harden, is rather low, it is cooled, after blazing, in water. The cooling after blazing thus being employed to equalize any slight difference in the heat of the spring when hardened.

The furnace is about ten inches wide and about four inches longer than the longest spring. The grate bars are arranged *across* the furnace with a distance of three-eighths of an inch between them.

The coal used is egg anthracite. It is first placed at the back of the furnace, and raked forward as it becomes ignited and burns clearly.

For shorter springs the coal is kept banked at the back of the furnace, so that the full length of the furnace is not operative, which, of course, saves fuel. By feeding the fire at the back end of the furnace, the gases formed before the coal burns quickly pass up the chimney without passing over the plates, which heat over a clear fire.

For commoner brands of steel, what is termed a water-chill temper is given. This process is not as good as oil-tempering, but serves excellently for the quality of steel on which it is employed. The process is as follows: The springs are heated and bent to shape on the *former* plate as before said; while at a clear red heat, and still held firmly to the *former* plate, water is poured from a dipper passed along the plate. The dipper is filled four or five times, according to the heat of the plate, which is cooled down to a low or very deep red. The cooling process on a plate 1 ½ x ¼ inches occupies about six seconds on an average, but longer if the steel was not at a clear red, and less if of a brighter red when the cooling began, this being left to the judgment of the operator.

Some brands of steel of the *Swede steel* class, will not temper by the water-chill process, while yet other brands will not harden in oil, in which case water is used to dip the plates in for hardening, the tempering being blazing in oil as described. In all cases, however, steel that will not harden in oil will not temper by the water-chill process.—*By* Joshua Rose, A. M.

TEMPERING SMALL TOOLS.

I have had a great many years' experience in the matter of tempering small tools, such as chisels, punches, drills, etc., and I think that I may be able, accordingly, to give some simple directions which will be of use to the trade. Steel for tools of this kind should never be heated beyond a bright cherry-red. The last hammering should be invariably on the flat sides, and at a very low heat; the tool meanwhile being held fairly on the anvil, and blows struck squarely with the hammer. This process is technically known as hammer-hardening, and serves to close the grain of the steel. After this, heat the article slowly and evenly until it shows a cherry-red about one and one-half inches from the point. Immerse the tool edge first into pure cold water, the surface of which is perfectly calm, a short distance, and hold it steadily at that point until the red above the water has become quite dull. Brighten the tool slightly by rubbing on a stone or by any other convenient means, and watch the changes of color which occur. First, blue will appear next to the body of the chisel, then probably brown, after which will come dark straw-color, then light straw-color, and at the edge a quite bright straw-color. Heat from the body of the tool will gradually change the location of these colors. The bright edge will assume a light straw-color; then will follow straw-color, orange, brown, and finally blue.

A word with reference to colors in the matter of tempering. A tool to cut hard cast-iron should be straw-color, or sometimes light straw-color; for soft cast-iron it should be dark straw-color; for wrought-iron, purple; or, if the iron is quite soft, blue may be found hard enough. A chisel treated as last described will wear well and not break easily. Drills may be tempered in the same way as described above. In my own practice, I leave them a shade harder than is required in a chisel. On the other hand punches may be slightly softer for the same work.—*By* Small Tools.

TO TEMPER SMALL PIECES OF STEEL.

Many blacksmiths are bothered to temper small pieces of steel on account of their springing. My way is to temper them in linseed oil, and they give me no trouble. I make a great many very thin knives.

Take a tin can large enough to insert the steel that you want to temper, say a tall two-quart fruit can, fill it with oil. Do not work your steel too hot, for that will spoil any steel. Temper just the same as you would if you were using water, and I think the result will be satisfactory.—*By* Volney Hess.

HARDENING THIN ARTICLES.

In hardening any article of steel that is thin or light and heats quickly, it is advisable to remove on a grindstone or emery wheel the scale formed in forging before heating. The scale being of unequal density, if it is not removed it is generally impossible to heat evenly; besides, the degree of heat can be better observed if it is removed.

SWORD BLADES.

Sword blades are made and tempered so that they will chip a piece out of a stone without showing a nick upon their edges, says a gentleman who has been through the great sword manufactory at Soligen, Germany. The steel, he says, is cut from bars into strips about two and one-half inches wide, and of the required length, by a heavy cutting machine. These are carried into the adjoining forge room, where each piece is heated white, hammered by steam so that about twenty blows fall upon every part of its surface, and then thrown into a barrel of water. Afterward these pieces are again heated in a great coke fire, and each goes through a set of rolls, which reduce it to something like the desired shape of the weapon. The rough margins are trimmed off the piece of steel in another machine, and there is left a piece of dirty, dark-blue metal shaped like a sword, and ready for grinding. This is done on great stones, revolved and watered by machinery, the workmen having to be the most expert that can be obtained, as the whole fate of the sword is in their hands. It is afterward burnished on small wheels managed by boys from twelve to sixteen years old, and when it has been prepared to receive the fittings of the handles, is ready for testing, which has to be done with great care. Any fault in the work is charged to the workman responsible for it, and he has to make it good. It is said that any blade which will not chip a piece out of a stone without showing a nick on itself is rejected.

TO TEMPER STEEL ON ONE EDGE.

Red-hot lead is an excellent thing in which to heat a long plate of steel that requires softening or tempering on one edge. The steel need only to be heated at the part required, and there is little danger of the metal warping or springing. By giving sufficient time, thick portions may be heated equally with thin parts. The ends of wire springs that are to be bent or riveted may be softened for that purpose by this process, after the springs have been hardened or tempered.

HEATING TO A CHERRY-RED—POINTS IN TEMPERING.

What is a cherry-red heat? To answer this question I will describe how I do my work. My shop is well lighted by windows and I heat to a cherry-red in the shade on days when the light is good, but on cloudy days I don't heat quite so high. I do not think that in hardening cast-steel it should be heated above a cherry-red in the shade. After hardening, temper thus: for razors, straw color: penknives, slightly bluish; screw taps and eyes, yellow; chipping chisels, brownish yellow; springs, dark purple; saws, dark blue.—*By* John M. Wright.

BRINE FOR TEMPERING.

Brine for tempering is usually known as hardening liquid, or hardening compound. The tempering is done after the hardening, and is usually termed drawing the temper. If good crucible steel is used there is nothing better than rain or soft water to harden with, that is, for tools which are not so thin as to bend, spring, or break when hardening. When thin articles, such as knife or saw blades, are to be hardened, cold raw linseed oil is the best material or compound extant. The bath may be filled entirely with oil, or have a surface of oil say six inches, and six inches or more of oil underneath the water. Clear oil is best, because, when water is present the lower stratum of oil is likely to saponify or become soapy, and loses its cooling qualities.

When the steel is not uniformly hard or lacks the necessary amount of carbon for hardening properly, then re-agents called hardening compounds are used to produce the necessary hardness. Of these compounds there are many, some of value, and more of questionable character. I give a few good preparations. Avoid using hard water at all times.

Chloride of sodium (salt), 4 ounces; nitrite of soda (saltpetre), ½ ounce; alum pulverized, 1 dram; soft or rain water, 1 gallon; when thoroughly dissolved heat to cherry-red and cool off. This process hardens and tempers, or draws no temper.

Another preparation is: Saltpetre, 2 ounces; sal-ammonia, 2 ounces; pulverized alum, 2 ounces; salt, 1 ½ pounds; soft water, 3 gallons. It is not necessary to draw with this mixture.

Another compound is: Corrosive sublimate, 1 ounce; salt, 8 ounces; soft water, 6 quarts. Corrosive sublimate is a subtle poison, so be careful with it.

While the above are of more or less value, the following will ever stand by you. It is in more general use than all the others put together:

Ferrocyanide of potassium, sometimes called prussiate of potash, 8 ounces, pulverized; 6 pounds of salt; 8 ounces of borax, pulverized; soft water, 10 gallons. The potash is poison, but when dissolved it becomes so well distributed in the water that its power as a poison becomes dissipated. For this mixture, as well as all others, use a wooden vessel, or a vitrified earthen vessel. The former is preferable. If prussiate of potash is not at hand, substitute 1 ½ pounds of crude potash, or two gallons full-strength soap-makers' lye, made by leaching hardwood ashes. In using this compound you will find that the salt and other ingredients are drawn to the surface edges of the vessel and form an incrustation on the outer upper section of the vessel, which you must remove and replace in the water. If in constant use replace the evaporated water with an equal amount pro rata of the ingredients. When much scale is deposited in the bottom of the vessel, remove the water to a clean receptacle, and clean the deposit from the vessel; replace the water drawn off and add sufficient water to replace the loss, and add ingredients pro rata. This preparation is in use by all file makers, and in it are hardened the smallest files extant. With care (keeping covered when not in use), a bath of this preparation may be kept with proper additions for years. I know of one bath of this kind which (by the necessary additions) has been in constant use thirty years. It is a tank holding about eighty gallons. Its owner would not dispose of it for $1,000. He makes the best files produced in America. For tools of any kind which are not liable to spring in hardening I do not know of a better ordinary process.—*By* Iron Doctor.

A BATH FOR HARDENING STEEL.

I have a hardening bath that is good and cheap. It is made as follows: Potassium cyanide, 2 ounces; ammonia carbon, 1 ounce; soda bicarbon, 1 ounce; aqua pura (water), 1 bbl.; sodium chloride (salt), 15 lbs.

I use a coal oil barrel. The plow lay should be of an even cherry-red all over. Hold the lay with tongs at the heel, put it into the water point first, but not too fast or it will spring, and keep the lay under water until it is cold.—*By* E. W. S.

THE LEAD BATH FOR TEMPERING.

Among the many secrets of tempering is the employment of the lead bath, which is simply a quantity of molten lead, contained in a suitable receptacle and kept hot over a fire. The uses of this bath are many. For instance, if it be desired to heat an article that is thick in one portion and thin in another, every mechanic knows how difficult it is to heat the thick portion without overheating the thin part. If the lead bath be made and kept at a red heat, no matter how thick the article may be, provided sufficient time be given, both the thick and thin parts will be evenly and equally heated, and at the same time get no hotter than the bath in which they are immersed.

For heating thin cutting blades, springs, surgical instruments, softening the tangs of tools, etc., this bath is unequaled.

If a portion of an article be required to be left soft, as the end of a spring that is to be bent or riveted, the entire spring may be tempered, and the end to be soft may be safely drawn in the lead bath to the lowest point that steel can be annealed without disturbing in the least the temper of the spring not plunged in the bath. Springs, or articles made of spring brass, may be treated in the same manner. A great advantage in the use of the lead bath is that there is no risk of breakage by the shrinkage of the metal at the water line, as is often the case when tempered by the method of heating and chilling in cold water.

As lead slowly oxidizes at a red heat, two methods may be used to prevent it. One is to cover the surface of the lead with a layer of fine charcoal or even common wood ashes. Another, and a better plan, is to float on the top of the lead a thin iron plate, fitting the vessel in which the lead is contained, but having a hole in the center or in one side, as most convenient, and large enough to readily admit of receiving the articles to be tempered or softened.—*By* W. H.

HARDENING SMALL TOOLS.

It is said that the engravers and watchmakers of Germany harden their tools in sealing wax. The tool is heated to whiteness and plunged into the wax, withdrawn after an instant and plunged in again, the process being repeated until the steel is too cold to enter the wax. The steel is said to become, after this process, almost as hard as the diamond, and when touched with a little oil or turpentine the tools are excellent for engraving, and also for piercing the hardest metals.

HARDENING IN OIL VS. HARDENING IN WATER.

I have made and tempered cutters for straight and irregular molders, sash tools, etc. If tempered in oil they will hold their edges better, cut smoother and longer than if tempered in water. In hardening, the oil cleaves to the steel, which is in consequence longer cooling. The water seems to separate, leaving air spaces between the steel and the water. Water cools quicker and hardens harder than oil and consequently steel hardened in oil must be left at a higher color than when hardened in water. A good deal depends on the heating of the steel to get a good temper. While water injures the quality of the steel, oil improves it.—*By* D. D.

TEMPERING PLOW POINTS.

If the steel is good, nothing will temper a plow point better than good clear water, with perhaps a little salt in it. Harden at as low a heat as the steel will bear, and do not draw the temper for blunt tools for cutting stone. Heat your plow points in the same manner. The great trouble with plow points is in the poor quality of the steel. You may make the plow points better by case-hardening, which every blacksmith knows how to do, and harden at a low heat without drawing the temper.—*Scientific American.*

TEMPERING BLACKSMITHS' TOOLS.

Some blacksmiths will, perhaps, be glad to know that by sifting prussiate of potash on red-hot iron and cooling it immediately, a temper is obtained hard enough to make a great many of the anvil tools used by smiths.—*By* I. C.

SOFTENING CHILLED CASTINGS.

For softening chilled castings my plan is as follows: Heat the metal you wish to drill in the fire to a little above a cherry-red. Then remove it from the fire and immediately place a lump of brimstone (sulphur) on the part to be drilled. You can keep the metal on the fire so as to retain heat and continue to throw sulphur on it until it will be as easy to drill as pot metal is.—*By* F. B.

TO HARDEN CAST-IRON.

Heat the iron to a cherry-red, then sprinkle on it cyanide of potassium (a deadly poison), and heat it to a little above red, then dip. The end of a rod that had been treated in this way could not be cut with a file. Upon breaking off a piece about half an inch long, it was found that the hardening had penetrated to the interior, upon which the file made no more impression than upon the surface. The cyanide may be used to case-harden wrought-iron.—*Scientific American.*

BRASS WIRE—HOW SHOULD IT BE TEMPERED FOR SPRINGS?

Brass wire cannot be tempered, if, by tempering, is meant the method of tempering steel springs. The only method to make a brass spring is by compressing the brass by means of rolls, or by hammering. The latter method will be the one that will probably be used. If the springs are to be flat, hammer them out to shape in thickness from soft wire, or sheet brass, somewhat thicker than the finished spring is to be. If the brass shows a tendency to crack in hammering it must be annealed, which can be done by heating to a light red and plunging into water. In hammering use a light hammer, and don't spare the blows.—*By* H.

TO HARDEN STEEL CULTIVATOR SHOVELS.

Shovel blades will harden in good soft water, with a little sal ammoniac in it, say one pound to ten gallons of water, if they are not put in too hot. Steel will harden best at a given heat, and practice alone will teach what that heat is. Clean the blades well, or, better still, polish them; then cover them with a paste made of salt and shorts, or flour and heat them very slowly.—*By* E. J. C.

CHAPTER V.

FORGING IRON.

HAND FORGINGS.

Notwithstanding the working of iron is one of the oldest mechanical occupations, the opinion prevails that the blacksmith need not of necessity be a man of much skill beyond what is necessary to heat and pound iron into shape. Yet this is but a small part of the duty required of him. If he is to work metal intelligently he must know the nature of the different kinds, its adaptability for specific uses. One grade of iron is well suited to one class of work and unfit for another. Then, too, welding is something more than the causing of two pieces of iron to adhere. Unless fusion is perfect the work is not well done, and perfect fusion can be obtained only by the proper heat and fit condition of the surfaces. Tempering steel is an important part of the business, and one which few thoroughly understand, or even care to.

In all other departments of carriage-making, patterns can be used with a certainty as guides, but the blacksmith has a large part of his work to do without patterns of any kind other than as general outlines for forms. Beyond this the eye must guide him, and as his work must be performed while the iron is hot, he must act promptly, and unless he is accurate his work must be done over again.

In ironing a carriage the woodwork is placed before him in pieces, and to perform his work well he must decide upon a general line of action before beginning his work. It will not do for him to make one piece and then take up another without previous study. If irons are to be fitted to the wood they must be shaped on the anvil, not on the wood, where, between burning and hammering, the general shape is obtained, but at the expense of the timber. Notwithstanding all that has been written and said against fitting iron to timber by burning, a large percentage of ironworkers in the country never think of any other way. We go into shops and see the blacksmith heat his tire red-hot, place it around the rim of the wheel, and amid fire and smoke "set up" to the rim, while two or three boys are busy with pails of water cooling off the iron. Those who have profited by experience and study find that if the tire is welded to the right length, and bent true, very little heat is required to expand the tire large enough to allow its being placed outside of the felloes, while water poured against the tread of the tire will cool it. In fact, not a few set light tires without using any water, and those who heat the tire but little find that the work is far better than when heated to a point that will burn the wood.

Too many blacksmiths are satisfied if the work is done so as to pass, regardless as to details and accuracy. These men are a disgrace to their vocation, and they too often regulate wages to be paid, and as the number increases more rapidly than that of the good workman, they will interfere the more seriously with wages in the future.

The young man who thinks of learning the blacksmith trade should first learn whether he is physically fitted for the peculiar labor. If satisfied on that point, he should immediately begin a course of study with special reference to the working of metals. He should also study free-hand drawing. Every hour spent at the drawing-board is an hour shaping irons, as he is training the hand to perform the work and the eye to see that it is true. And at no time should he drop the pencil. He should keep in mind the fact that the most skillful are the most successful. We do not mean skillful in one line only, but in all. The man who can direct, as well as execute, is the one who will make the greatest advancement, and to direct it is necessary to know why a thing should be done as well as how.

The poetry of the blacksmith shop has been a theme for writers for centuries, but there is little poetry in it to the blacksmith who stands at the forge day after day pounding and shaping unless he has studied, and finds new themes in every heat, spark, or scale. If he can create beautiful forms in his mind, and with his hands shape the metal to those forms, then he can see poetry in his work. If he is but a machine that performs his work automatically, the dull prose of his occupation makes him dissatisfied and unmanly.—*Coach, Harness and Saddlery.*

MAKING A T-SHAPED IRON.

If the blacksmith who wants to learn how to make a T-shaped iron, will take good iron two-thirds as long as his head block, twice as wide and one-eighth of an inch thicker than he wants the plate when finished (or perhaps one and one-sixteenth inches on a light plate), fuller in one edge one-third from the end, draw off the short end, next split the long end down to within one and one-quarter inches of the fuller scarf (if he has an inch perch), then rest the fuller scarf on the round edge of the anvil, follow the chisel down with the small fuller and punch, and turn off to a right angle with a hand hammer, his iron will be so nearly finished that he will understand what else should be done without any more directions from me.

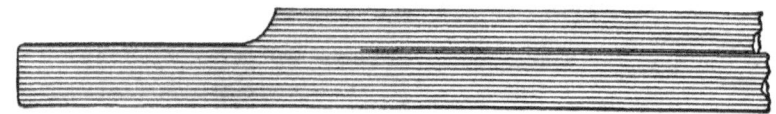

MAKING A T-SHAPED IRON. FIG. 128—SHOWING THE IRON READY FOR TURNING.

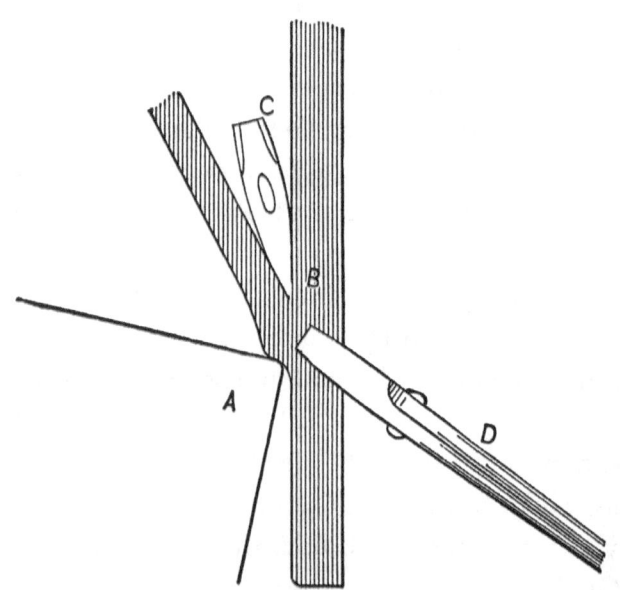

FIG. 129—SHOWING THE TURNING PROCESS.

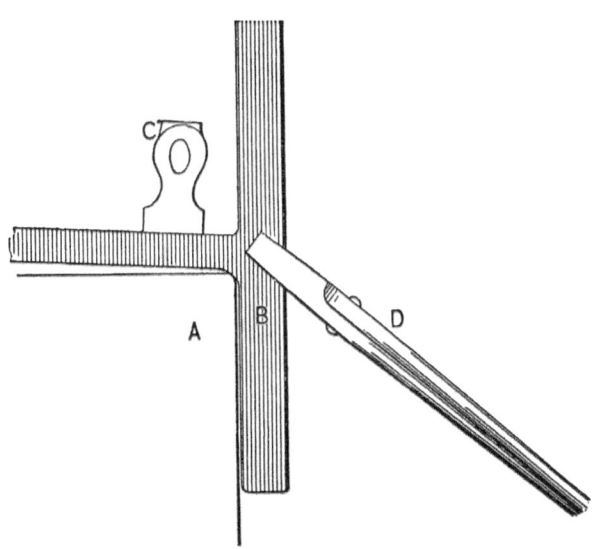

FIG. 130—SHOWING HOW THE IRON IS TURNED TO A RIGHT ANGLE.

The accompanying illustrations will help him to understand my instructions. Fig. 128 represents the iron ready for turning; Fig. 129 shows the turning process, *A* denoting the anvil, *B* the iron, *C* the fuller, and D the tongs. In Fig. 130, A denotes the anvil, *B* the iron, *C* the hammer, and *D* the tongs.—*By* E. K. W.

ANOTHER METHOD OF MAKING A T-SHAPED IRON.

I will describe my way of making a T-shaped iron without having a weld at the hole that the kingbolt goes through.

Take a piece of good iron of suitable size for the job you want and fuller and draw out the end as at Fig. 131. Next place B on the anvil and insert the fuller on the two inside corners so as to draw out the ends for the head block, then weld to each end to make it the desired length.

Do not cut the inside corners square, but leave them rounding, as Fig. 132, and also a trifle thicker at the center, as this makes it stronger and at the same time gives more support to the kingbolt. I almost always weld at *A*, Fig. 132, by upsetting well, so as to have plenty of stock, and being careful to get a clean heat. I do not remember ever having one break that was made in this way.—*By* J. L.

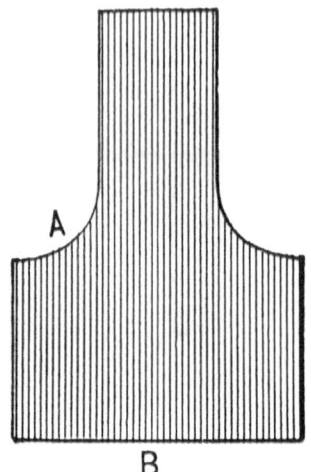

MAKING A T-SHAPED IRON BY THE METHOD OF "J. L."
FIG. 131—SHOWING HOW THE END OF THE IRON IS FULLERED AND DRAWN OUT.

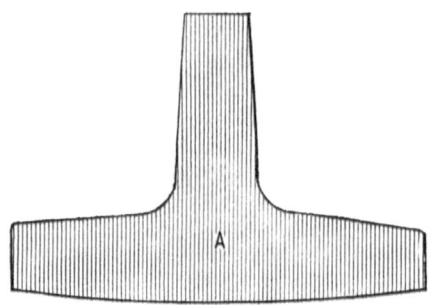

FIG. 132—SHOWING THE T COMPLETED.

FORGING STAY ENDS AND OffSETS.

To make stay ends I take a piece of good iron 1 ½ x 1 inch and fuller and forge as shown in the dotted lines of Fig. 133.

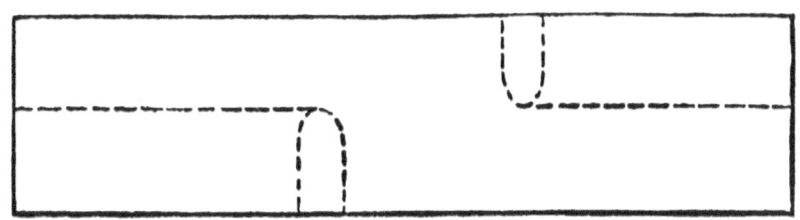

FIG. 133—SHOWING THE STAY END WITH DOTTED LINES FOR FORGING AND FULLERING.

It will then be shaped as in Fig. 134. I then fuller and forge as in the dotted lines of Fig. 134, and finish with the file. To make a stay offset take a piece of iron 1 ½ x ¾ inch, forge as in dotted lines of Fig. 135, punch a hole and split open. Next open as in Fig. 136, and forge as seen in the dotted lines, then straighten up and finish with a file.

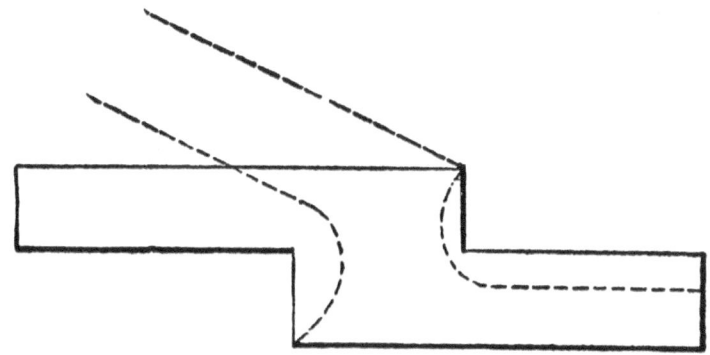

FIG. 134—SHOWING THE PIECE AS FORGED AND FULLERED.

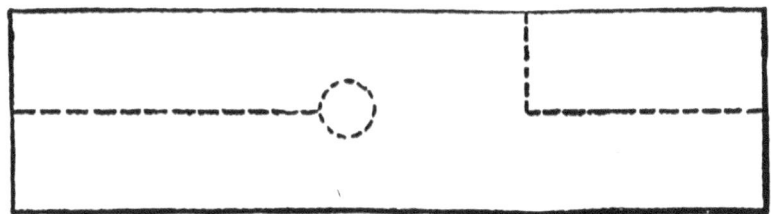

FIG. 135—SHOWING THE STAY OFFSET WITH DOTTED LINES FOR FORGING.

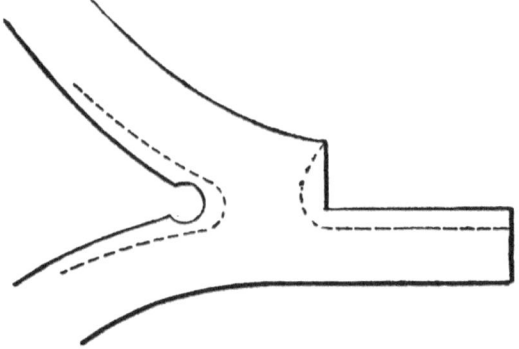

FIG. 136—STAY OFFSET OPENED TO FORGE AS IN DOTTED LINES.

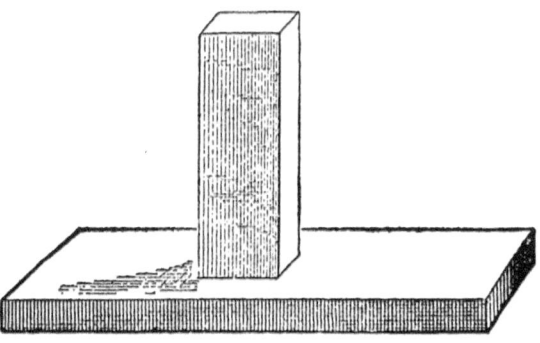

FIG. 137—SHOWING ANOTHER METHOD OF MAKING STAY ENDS.

Figs. 137, 138 and 139 illustrate another way of making stay ends.

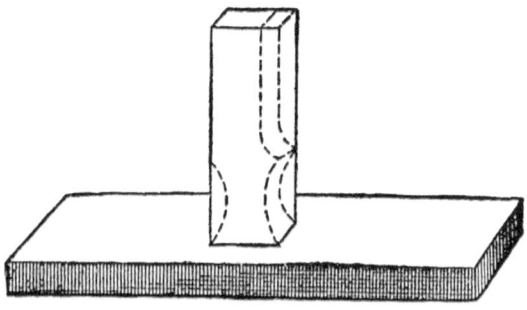

FIG. 138—SHOWING THE FULLERING AND FORGING AS BY DOTTED LINES.

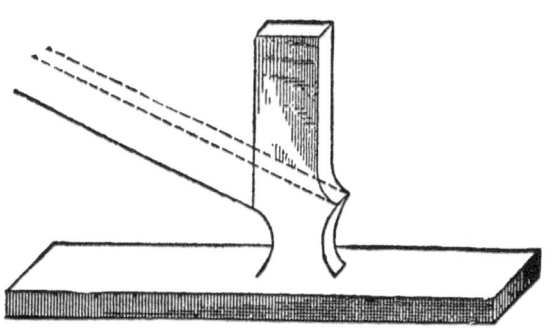

FIG. 139—SHOWING THE SHAPE AFTER FULLERING AND FORGING.

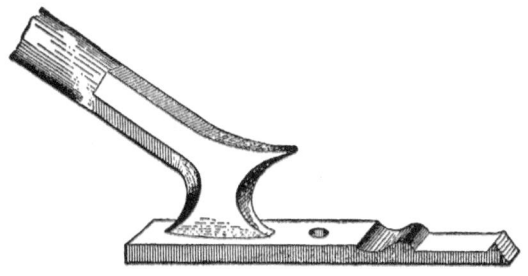

FIG. 140—SHOWING FINISHED STAY END.

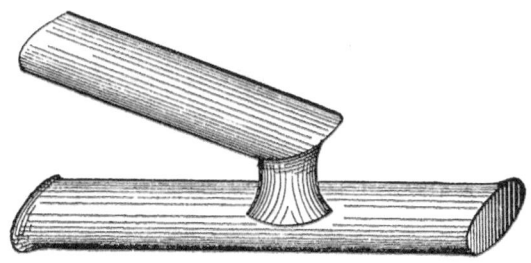

FIG. 141—SHOWING FINISHED OFFSET.

Forge a piece of iron, square and small enough to go in the square hole of the anvil; then forge on the flat head, giving a shape as in Fig. 137. Next fuller out and forge as shown in the dotted lines of Fig. 138. This will give a shape as in Fig. 139; then bend according to the dotted lines in that figure, and finish with a file. Figs. 140 and 141 show the finished end and offset.—*By* J. C. H.

MAKING AN EYE-BOLT.

The accompanying illustrations represent my method of making an eye-bolt so that it will be round in the eye, and likewise very strong at the weld.

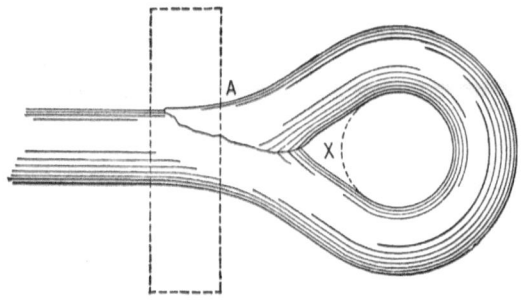

MAKING AN EYE-BOLT. FIG. 142 — SHOWING THE OLD WAY OF DOING THE JOB.

Fig. 142 represents the old way of making an eye-bolt and the way it is made now by most blacksmiths.

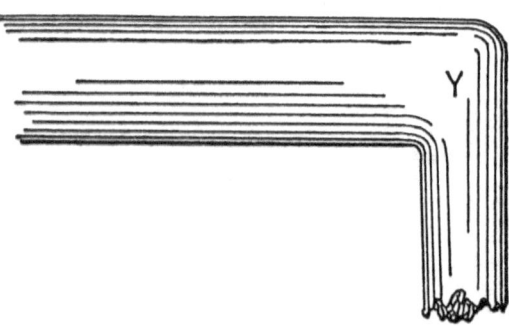

FIG. 143—SHOWING HOW THE IRON IS TURNED IN E. K. WEHRY'S METHOD OF MAKING AN EYE-BOLT.

The bolt is simply turned about, the end weld being at A. It will be noticed that the place X is not filled out with iron, and that the hole in the bolt cannot be round unless more iron is used.

By my method the iron is turned down as at Y in Fig. 143. It is made somewhat flat at Y and then turned around as in Fig. 144.

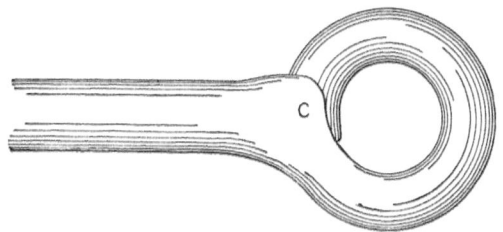

FIG. 144—SHOWING HOW THE WELD IS MADE.

A good weld is made at *C*, and it is worked down to the size of the iron. The eye-bolt is as shown in Fig. 145. The weld will not give and the hole is a round one. With a little practice, this eye-bolt can be made as easily as the old style, and can be worked down so that it will fit in a hole up to the eye as shown in Fig. 145 at *K*.

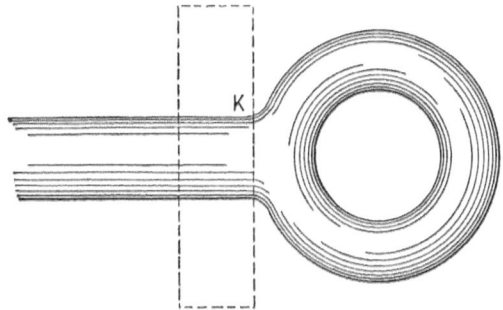

FIG. 145—SHOWING THE EYE BOLT COMPLETED.

Compare this fit with that shown in Fig. 140.—*By* E. K.

FORGING A TURN-BUCKLE.

In forging a turn-buckle, I first make a mandrel, as shown in Fig. 146.

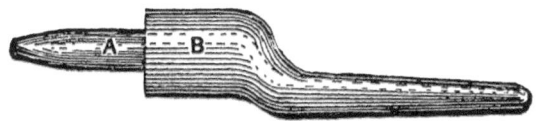

"TINKER'S" METHOD OF FORGING A TURN-BUCKLE. FIG. 146—
SHOWING THE MANDREL.

The part *A* is one and one-fourth inch round, *B* is one and three-fourths inch square. I next take some 1 ¼ x 1 ½ inch iron, make collars and weld them on the mandrel, and scarf on each side of the collars, with a round pene fuller for the check pieces, as shown in Fig. 147. I then take for the sides some ⅞ x 1 ⅛ iron and cut long enough for both sides, bend in the middle and scarf the ends, and am now ready for welding up. I put one of the collars between the two prongs and take a light heat to stick them together.

 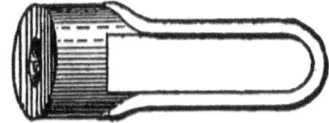

FIG. 147—SHOWING THE FINISHED BUCKLE AND COLLAR.

I then go back to the fire and get a good soft heat, weld down on the mandrel, finish off with the swage, then cut the end and repeat the process. If the turn-buckle is to be finished up extra nice, you can use the same swage for the sides.—*By* "Tinker."

My method of forging a turn-buckle, say for a one and one-fourth inch rod, is: First, I take a piece of one and one-fourth inch square iron and bend it into a ferrule such as is shown in Fig. 148 and then weld into it a prong as in Fig. 149.

FORGING A TURN-BUCKLE.
FIG. 148—END VIEW OF FERRULE BENT FOR WELDING.

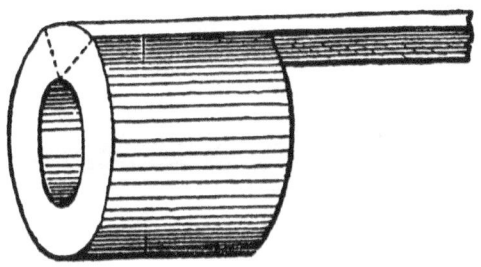

FIG. 149—SHOWING PRONG WELDED TO FERRULE.

Then I cut out a place, as at *A* in Fig. 150, and weld in the other prong. Next I put in a mandrel and shape it up to the taper, as in Fig. 151. This makes half the turn-buckle and by repeating the operation I get two of these pieces, which I weld together at *B B*, and get the finished turn-buckle as in Fig. 152.

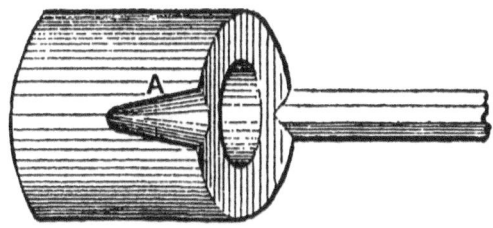

FIG. 150—PIECE CUT OUT AT A FOR SECOND PRONG.

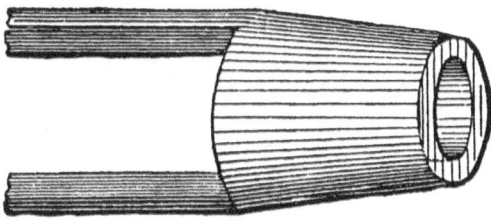

FIG. 151—SHOWING SECOND PRONG WELDED AND TAPER FORGED.

In welding the prongs together to join the two halves, I take one-half and let my helper put in the mandrel with one hand and tap lightly with a small hammer with the other, and after the welding is done I use the mandrel to handle the forging with while swaging it to finish.—*By* Southern Blacksmith.

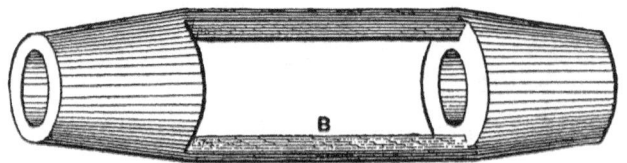

FIG. 152—SHOWING THE TWO PIECES WELDED AT B B, AND JOB COMPLETE.

MAKING A CANT-HOOK.

I make a great many cant-hooks. For the clasp I use 1 ½ 1 ½ inch Norway iron. About half an inch from the end I cut it half off, bend over and weld, forging rather thin inside of the shoulder. I then take a set hammer and make the shoulder for the hook to strike against.

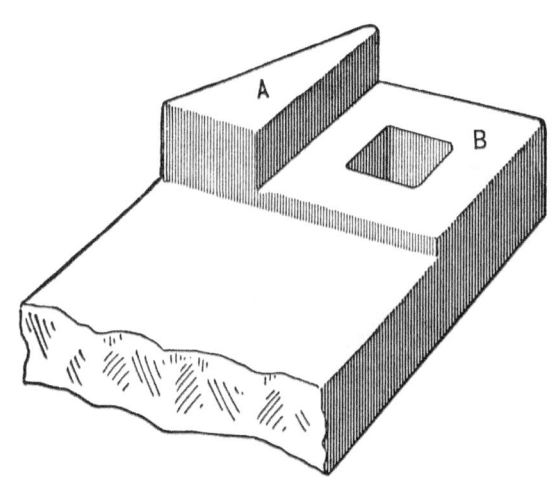

MAKING A CANT-HOOK.
FIG. 153—SHOWING HOW THE JAM AND CLASP ARE MADE.

I leave the jaw about three-eighths inch thick, punch the hole, and trim to suit. The other end is made in the same way. For the hook I use ¾ x ½ x ⅞ x ½ inch steel, cut thirteen inches long. One end should be stove (not bent) and put in the heating tool to get the bill. After taking it out I draw it to a point, punch the hole and bend nearly to a circle. In bending I lay the bill and eye on a straightedge or board as shown in Fig. 154.

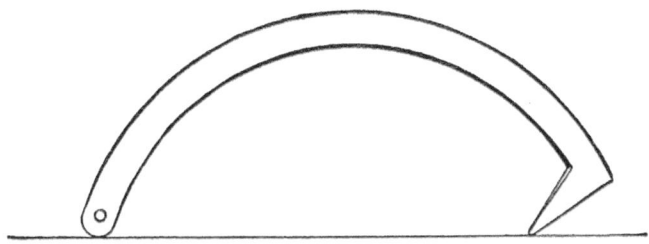

FIG. 154—SHOWING HOW THE BILL AND EYE ARE BENT.

I use three bands, the first and second being one and one-half inch No. 14 band iron; the third band is two and one-half inch No. 8.

For the pick I use seven-eighths inch square steel, and cut it off ten inches long.

In Fig. 153 of the accompanying illustrations *A* indicates the thickest part of the clasp, and *B* is the jaw. This is generally made too thin. The point must not come down far enough to touch the toe ring, but should stand up six inches. In bending the bill there is no sharp corner made which may break the hip when you are pulling it out of a log.—*By* H. R.

HOW FORKS ARE FORGED.

The following is a description of the manner of forging a four-tined manure fork, as practiced in one of the largest establishments in the neighborhood of New York City. The process of splitting and bending as here described may be extended so as to include forks of a larger number of tines.

HOW FORKS ARE FORGED. FIG. 155—THE BLANK.

The blank for a four-tined manure fork is simply a rectangular piece of mild steel, as indicated in Fig. 155.

FIG. 156—FIRST OPERATION, PREPARING THE HANDLE STEM.

It is five and three-quarter inches long, one and three-quarter inches wide, and one-half inch thick. The first operation is to draw out the end, leaving a projection, A, as shown in Fig. 156, from which the stem for the handle is subsequently drawn.

The piece is then split to the line B, Fig. 157, and is opened out as in Fig. 158. The thickness is reduced to C. All this is done at one heat, and the splitting and opening out is accomplished by a machine termed a splitter. The next operation is to split the piece along the lines $D\,D$ in the engraving.

FIG. 157—SPLITTING THE PIECE PREPARATORY TO OPENING IT, AS SHOWN IN THE NEXT ILLUSTRATION.

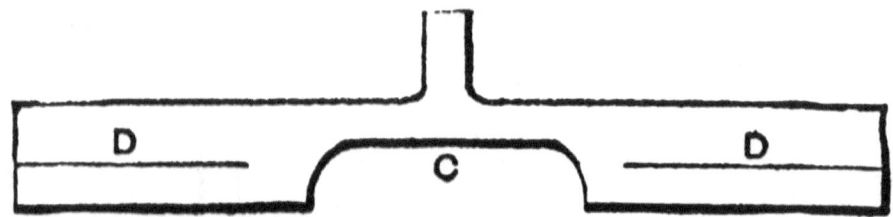

FIG. 158—THE PIECE OPENED OUT, FORGED DOWN AND SPLIT AGAIN.

The parts are then opened out as in Fig. 159, *EE*, forming, when forged out as in Fig. 160, the middle tines.

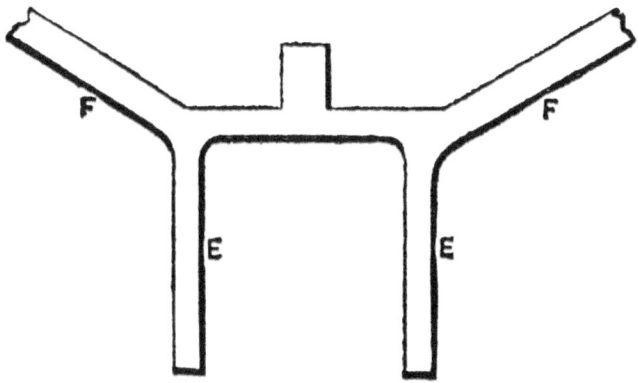

FIG. 159—THE FOUR TINES ROUGHLY FORMED.

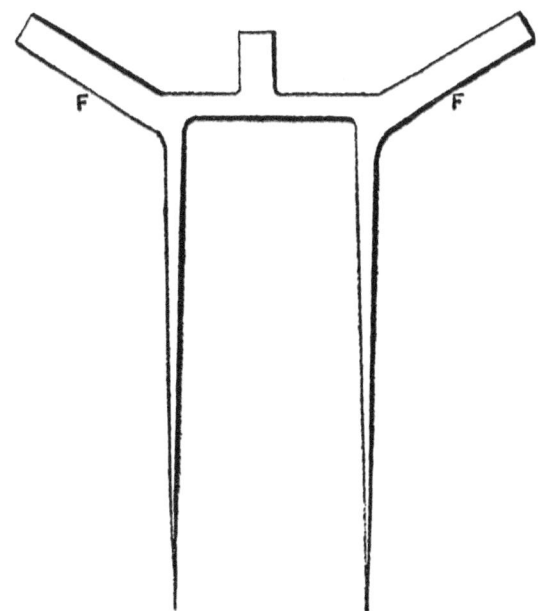

FIG. 160—THE MIDDLE TINES DRAWN OUT.

After the tines *EE* are drawn out, the ends *FF* are drawn and bent around as shown in Fig. 161. The last operation is to draw out the handle *A* to the required shape. After these several steps have been taken, the tines are formed to proper shape and finally are tempered.

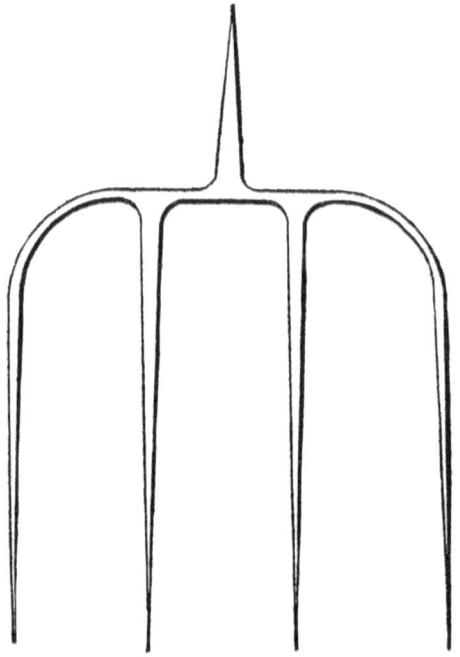

FIG. 161—THE FINISHED FORK.

It will be obvious to the practical reader that the grain of the steel in this process of forming the fork is kept lengthwise of the handle as well as of the tines. The whole of the forging operations described are performed under the trip hammer, but the forming or setting to shape of the tines is done in a special machine.

FIVE METHODS OF MAKING ONE FORGING.

The sketches given herewith are of a piece of forging of a somewhat intricate character. The different methods of accomplishing the same result form an interesting study. Fig. 162 shows the finished article.

To make this, says one blacksmith, of ordinary refined iron, I fuller it as at *A A*, and punch the hole at *B*, Fig. 163.

Then I split it open, as shown by the dotted line in Fig. 164, and open it out as in Fig. 165.

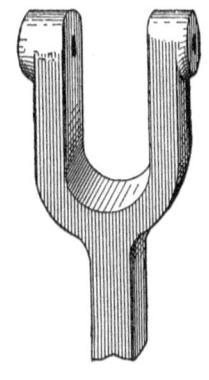

FIG. 162—THE FINISHED FORGING.

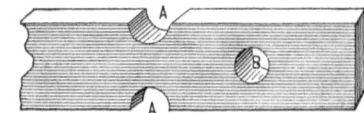

Fig. 163.

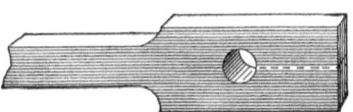

FIG. 164.

Then I bend each arm, as C D in Fig. 165, to shape and forge it to size, punching the holes toward the last.

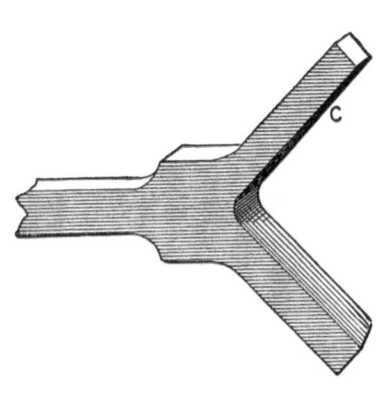

FIG. 165.

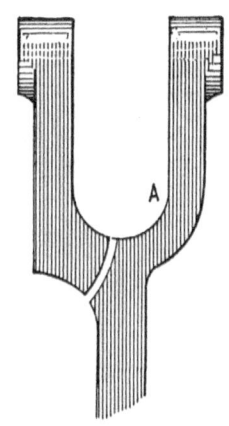

FIG. 166.

I know of no better way of forging it out of solid iron unless I could get the job roughed out to any better shape by sending it to a shop having a trip hammer to rough the blocks out. Or it might be a good plan to use lighter iron, forge the piece in halves and weld them as shown in Fig. 166, as my principal difficulty is in setting the jaw back for the curve *A* in Fig. 166.—*By Young Blacksmith.*

Method No. 2.

Take a piece of iron large enough, and bend it to U shape, as at *A*, Fig. 167. Cut the end as at mark *B*, and bend over and weld to make eyes, *D*.

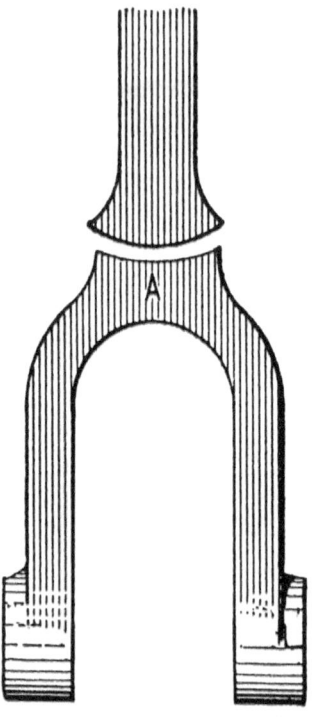

FIG. 167.

Then jump-weld the rod *E*. It will make the smoothest job to jump it, if the iron is thick enough.—*By* H. B.

Method No. 3.

Thinking it a duty, as well as a pleasure, to help my brother blacksmiths, I give you my way of doing the job.

I should take iron of sufficient size to forge the piece A, Fig. 168, fullering it in the center, as at a, with a large fuller to sufficient depth to make a good weld.

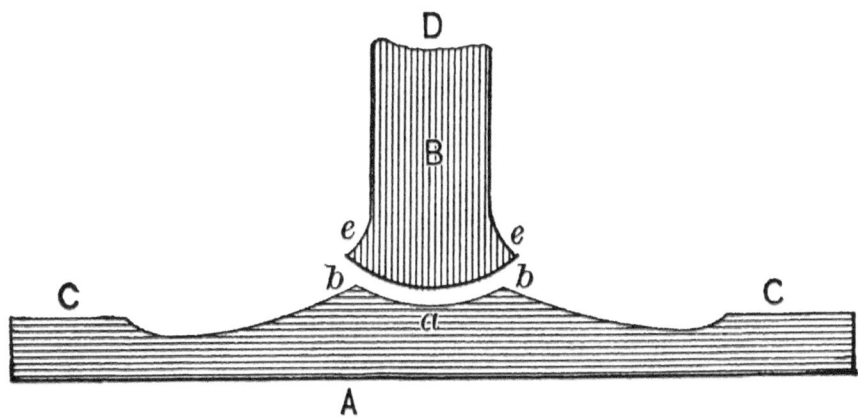

FIG. 168.

Then, with a small one, fuller it out on each side, as at *bb*, and draw it out long enough to bend to the required shape, leaving a lump, *CC*, on each end to form the eyes. Then forge the stem *B*, making the scarf at *ee*; overlap the fullering at *a*, and more curved than the fullering at *a*, so that when the jump-weld is made, the stem *B* will first meet the piece *A* at the bottom of the fullering at *a*, and the rest of the surface will come together with the first jumping blows. This is quite important in obtaining a good, sound weld, as it forces out the air and any loose cinders, etc., that may not have been cleaned off the surfaces. After welding at the center by striking on the end, *D*, of the stem, weld with a fuller applied at *ee*, and then bend to the required shape.

There is another and similar way, as follows: Form the two pieces as in Fig. 169, leaving it thick enough to weld and finish.

In this case, also, the fullering on the stem must be more flat than the seat on which it welds, so that it will be sure to weld in the center first, and weld on the horn of the anvil.—*By* G. B. J.

Method No. 4.

First take a piece of iron thick enough to give plenty of iron when fullered at *A* in Fig. 170, and fuller it hollow there. Then bend it to shape; take a piece for the stem and jump it, taking care that it well laps the fullering at *A*, and let it be more curved than at *A,* so that it will weld in the center first, To weld, place the hook on the anvil horn, clean it with a brush; bring the stem down on the anvil with a light blow to clean off the dirt, etc.; brush it quickly, strike a few quick blows with a light sledge on the end of the stem and then fuller round the stem.

Get a good heat and don't lose a second's time and you'll have a good, sound, neat job.

Some blacksmiths make the weld round at A instead of hollow, but I don't see why.—*By* Leather Apron.

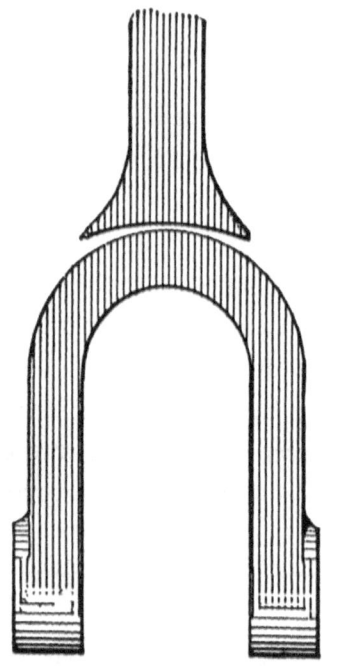

FIG. 169—METHOD NO. 3.

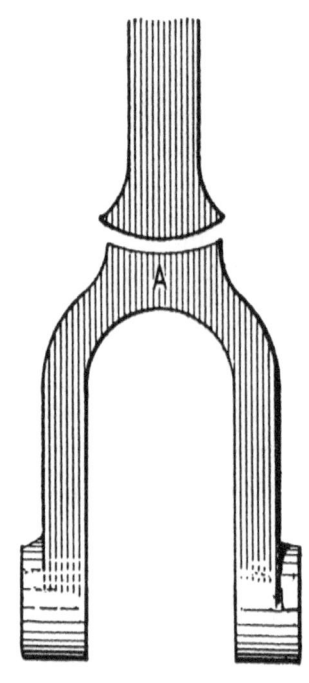

FIG. 170—METHOD NO. 4.

Method No. 5.

I would like to have blacksmiths try my method, as per Fig. 171. Forge two separate pieces, as *B, C,* in the figure, and weld them at *D.* Fuller *A A* to dotted lines, and draw out the stem.—*By* C. A. S.

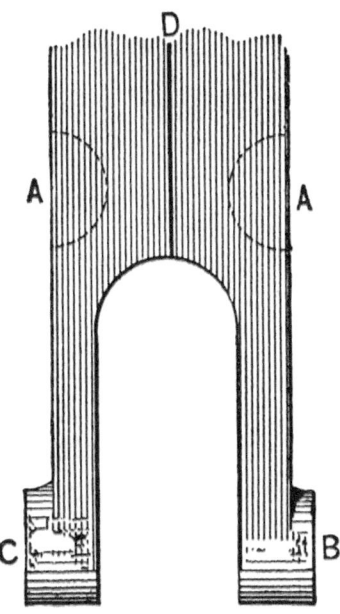

FIG. 171—METHOD NO. 5.

MAKING OFFSETS.

Take a piece of good iron of the proper size and fuller in half way at *A*, in Fig. 172, then swage down the end *B*, then split with a sharp chisel at *C*, down to *D*.

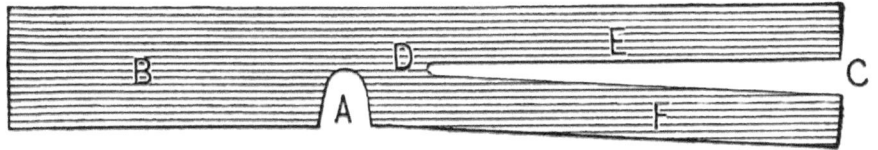

MAKING OFFSETS. FIG. 172—SHOWING HOW THE IRON IS FULLERED AND SPLIT.

Next with a small fuller work in at *D* and finish with a large fuller, then turn off the ends *E* and *F*, and swage down to proper size as shown in Fig. 173.—*By* J. D.

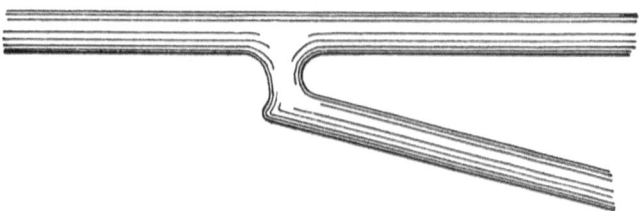

FIG. 173—SHOWING THE OFFSET COMPLETED.

TO MAKE A SQUARE CORNER.

It requires years of hard study and practice to attain any high rank in our profession, but if brother blacksmiths would each make known a little of what they have learned by practical experience, it would be a great benefit to the trade.

I have never met any man who had nothing to learn in blacksmithing, and yet I have known some very good workmen.

The illustrations herewith show a good way of making a square corner, with a fillet inside if required. The upset and scarf are as seen in Fig. 174, and Fig. 175 represents the corner complete.

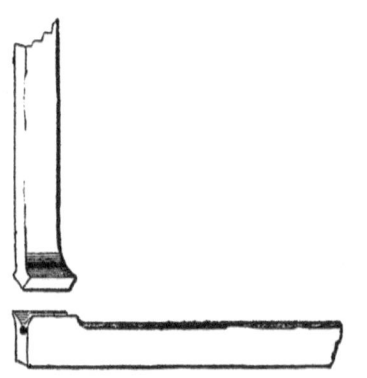

MAKING A SQUARE CORNER.
FIG. 174—READY FOR THE WELD.

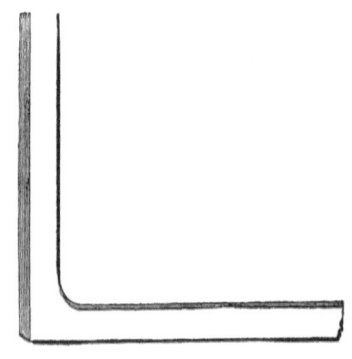

FIG. 175—COMPLETED JOB.

This works well in heavy iron, say two inches by four inches. It is a V weld in the corner. Be careful to have good clean heats.—*By* R. C. S.

MAKING A SQUARE CORNER.

I send you herewith my method of getting a square corner up sharp. In the accompanying illustration I show in Fig. 176 a piece of iron to be bent to a right angle and have a square corner.

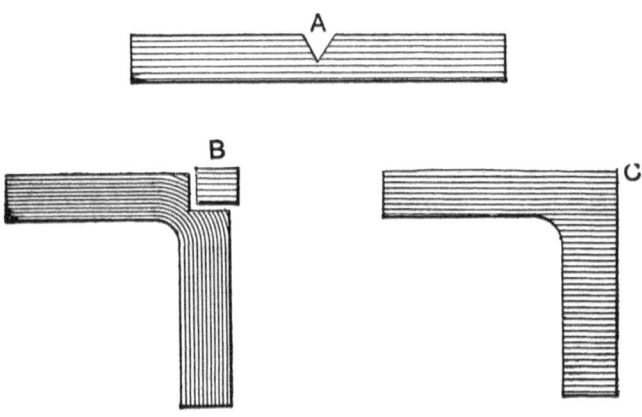

FIGS. 176, 177, AND 178—"SOUTHERN BLACKSMITH'S" METHOD OF MAKING A SQUARE CORNER.

First I cut it half way through where the bend is to be, as at *A*, then I bend it as shown in Fig. 177, and then weld in a piece *B*, which brings the corner up sharp and square as shown in Fig. 178.—*By* Southern Blacksmith.

THE BREAKING OF STEP-LEGS.

Builders of heavy wagons with the step-leg at the back end of the wagon experience much trouble from their breaking. They try to overcome this by making them of thicker and wider iron every year, but they invariably break in the outer hole, and frequently break off the bolts.

In making iron of such a character, at all times avoid, if possible, having holes where the direct strain comes. Fig. 179 shows the proper way to do such work. *A* is the upper part of the leg and fastened on the underside of the tail-bar. *B* is a projection, which is secured to the central sills. *C* is the shank of the leg. *D, D, D* are the bolt holes. An iron step made and secured in this manner with one or two central braces will last until the wagon is worn out, and will then answer for another wagon.—*By* Iron Doctor.

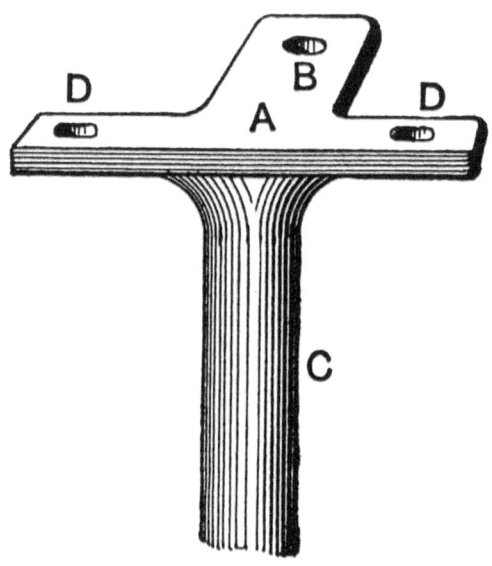

FIG. 179—SHOWING "IRON DOCTOR'S" METHOD OF PREVENTING THE BREAKING OF STEP-LEGS.

MAKING A THILL IRON.

My way of making a thill iron is to take a straight bar of iron of the proper width and bend it square around, forming the stem of the T, as shown by Fig. 180.

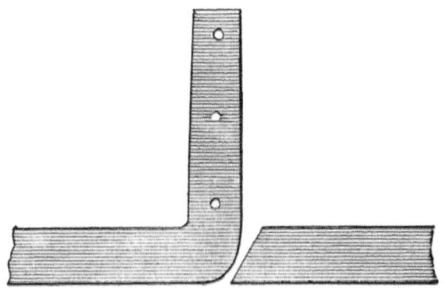

FIG. 180—MAKING A THILL IRON BY THE METHOD OF JOHN ZECK.

I then take another piece of iron of the same size as the first, cut the end to be welded a little slanting and weld it on at the bend of the first piece with the long corner out, as shown in the illustration. This leaves the iron straight on the outside, and when made in this way thill irons never break, as they often do when made in the ordinary way.—*By* John Zeck.

MAKING A T-SHAPED IRON.

To make a T-shaped iron, I always take the best of iron, Norway or Swedish, for material.

In making a T-shaped, or in other words head-block plates for a one-seated rig, I take a piece of Norway iron 3 x 3 x ¾ inches thick, heat the whole piece at once, put it on the anvil, and cut with a hot chisel as in Fig. 181, so that the prong will be 1 x ¾ inch. I next cut away a little of the corner, as shown by dotted lines in Fig. 182.

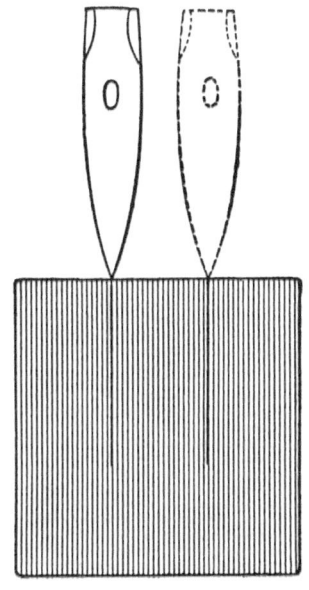
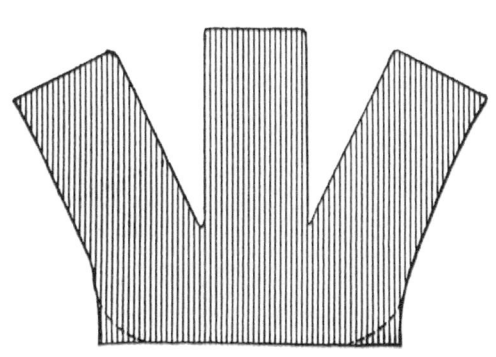

MAKING A T-SHAPED IRON. FIG. 181—SHOWING HOW THE PIECE IS CUT.

FIG. 182—SHOWING HOW THE CORNERS ARE CUT.

This prevents a seam, and doubtless a bad job. With two heats I bend with the fuller as in Fig. 183, and then draw out according to the shape of the head block, which must govern the thickness and width of the iron. Fig. 184 represents the piece after the fullering.

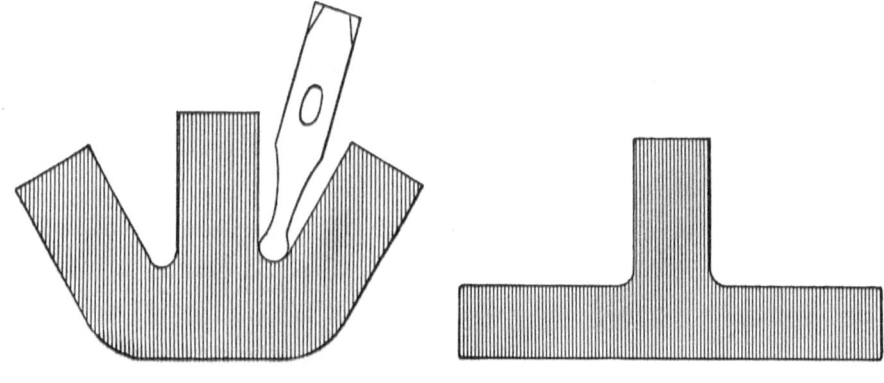

FIG. 183—SHOWING HOW THE FULLER IS USED.

FIG. 184—SHOWING THE PIECE AFTER BEING FULLERED.

For larger T's a larger piece of iron is needed. Be sure you have enough iron, as it will be far better to cut off than to weld on. Leave the corner rounded, as it adds strength.

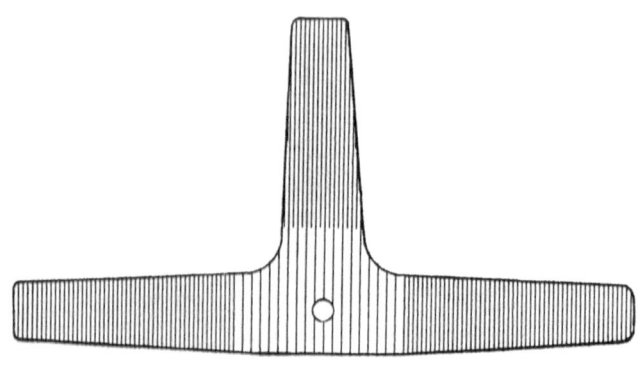

FIG. 185—THE FINISHED T.

Fig. 185 shows the T finished. After one or two trials you can make one in from three to five heats.—*By* L. A. B.

MAKING A STEP-LEG.

Take iron of the size that will fit in the square hole of the anvil, and split it as at *A* in Fig. 186, then bend it over nearly at right angles with the standing part, take a welding heat and beat down as in Fig. 187, *BB* being the upper part and *C* the leg.

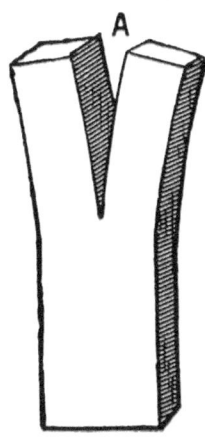

MAKING A STEP-LEG. FIG. 186—SHOWING HOW THE IRON IS SPLIT.

Then prepare the pieces *F*, Fig. 188, heat both to the welding point, insert the part *E* in the hole and weld *F* to *D*. If the square hole in the anvil is not countersunk, make tools as in Figs. 189 and 190. *H*, Fig. 189, is a hole, *L* and *B*, Fig. 190 are recesses, and *M* is a hole.

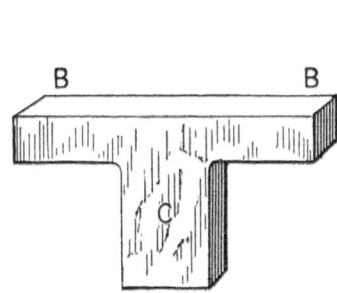

FIG. 187—THE IRON AS BENT AND BEATEN DOWN.

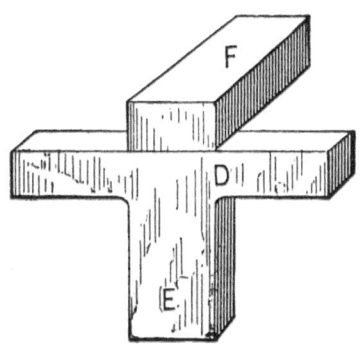

FIG. 188—SHOWING HOW THE PIECES F AND D ARE WELDED TOGETHER.

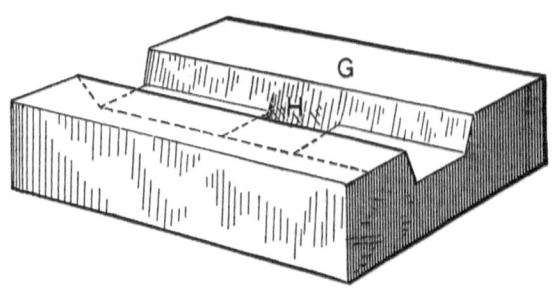

FIG. 189—SHOWING A TOOL USED WHEN THE SQUARE HOLE IN THE ANVIL IS NOT COUNTERSUNK.

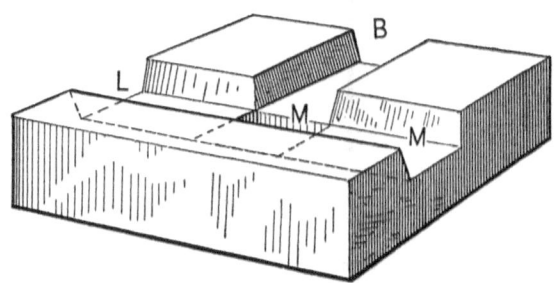

FIG. 190—ANOTHER ILLUSTRATION OF THE TOOLS USED IN MAKING A STEP-LEG.

The recesses may be rounded or flat on the bottom, and round at sides as in Fig. 191 or flat on the bottom and beveled on the sides as in Fig. 192.

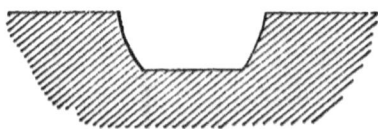 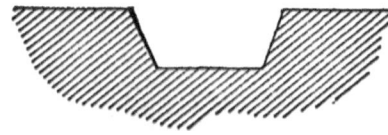

FIG. 191—SHOWING ONE METHOD OF SHAPING THE RECESS IN THE TOOL.

FIG. 192—ANOTHER WAY OF SHAPING THE RECESS.

Beveling or rounding are necessary to ensure the ready removal of the iron from the tool. The holes ought to be slightly countersunk to prevent galling.—*By* Iron Doctor.

FORGING A HEAD-BLOCK PLATE FOR A DOUBLE PERCH.

My method of forging a head-block plate for a double perch may be of interest to the trade. It is as follows:

Take good Norway iron 2 ½ x ⅜ inches, and forge as shown in Fig. 193, on the back of the anvil; then cut out as indicated by the dotted lines, Fig. 193, then split the ends and turn down for perch ends, finish in the corners with files, draw out the ends of the plate, and you have a forging for a double saddle clip as shown in Fig. 194.

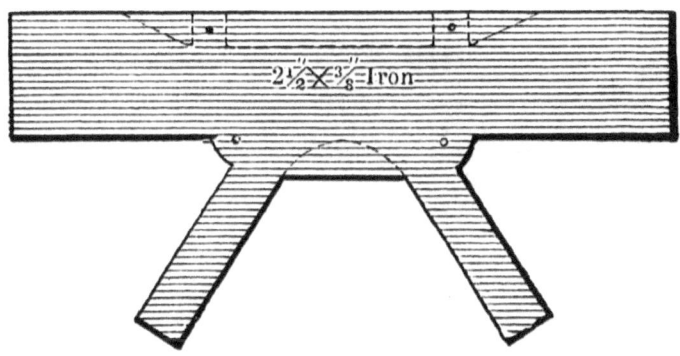

FORGING A HEAD-BLOCK PLATE FOR A DOUBLE PERCH.
FIG. 193—SHOWING HOW THE IRON IS FORGED AND CUT OUT.

To make a Brewster saddle clip take iron the same size as for the double saddle clip, and use the fuller as indicated in Fig. 195, then cut out according to the dotted lines, split the ends as in Fig. 193, and finish. Then you have a forging as shown in Fig. 196.

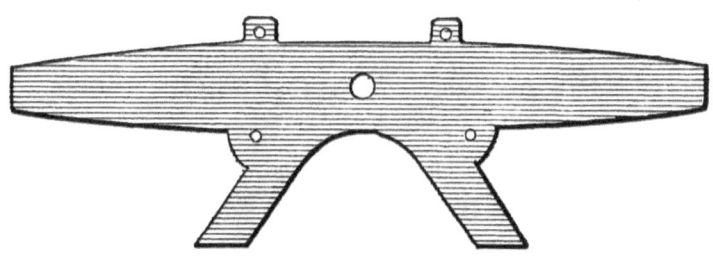

FIG. 194—SHOWING HOW THE FORGING IS DONE
FOR A DOUBLE SADDLE CLIP.

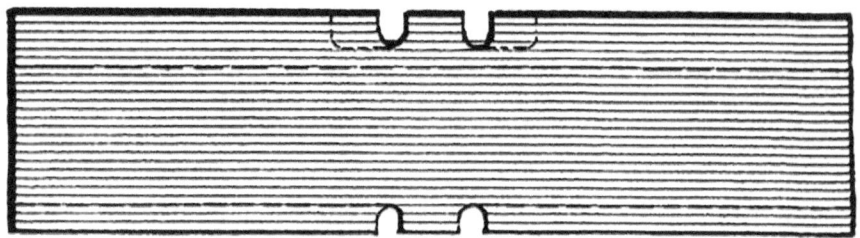

FIG. 195—SHOWING HOW THE FULLER IS USED IN MAKING A BREWSTER SADDLE CLIP.

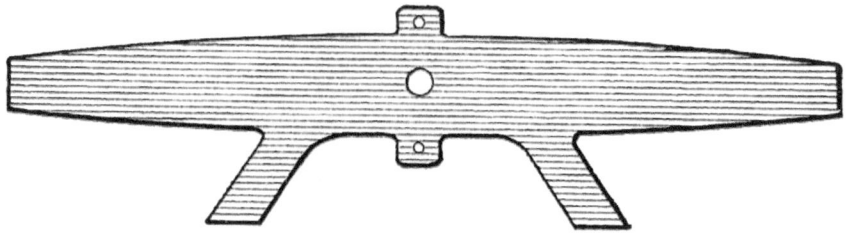

FIG. 196—SHOWING THE FORGING COMPLETED.

I don't claim that this method is cheaper or better than buying the machine-made articles, but if you want one immediately and cannot get it without waiting two or three days, the next best thing to do is to make it as I have described.—*By* P. R.

FORGING A DASH FOOT.

My way of forging a dash foot is as follows: I take a piece of Norway iron 1 ½ x 1 ½ inch and five or six inches long, fuller it at *A* as shown in Fig. 197, then forge out as indicated by the dotted line, punch a small hole at *B*, and then split out as shown by the dotted line.

The piece is then ready to be opened out as in Fig. 198.

The corners and ends are then squared by swaging and the rough places are finished with a file. The result is a dash foot as neat and substantial as any one could desire. The welding on is done at *C* and *D*.—*By* J. C. H.

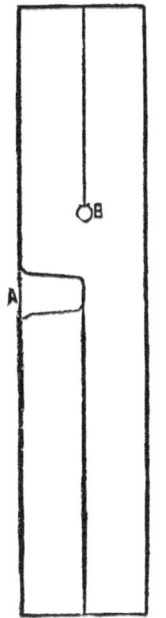 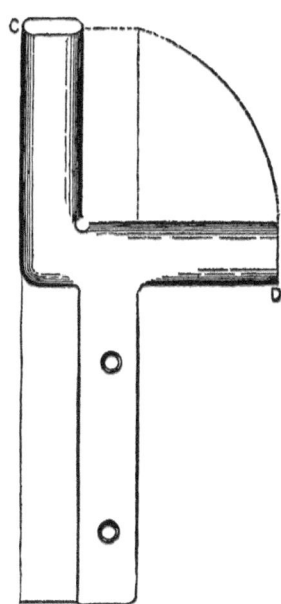

FORGING A DASH FOOT BY "J. C. H."
FIG. 197—SHOWING HOW THE IRON IS FULLERED, PUNCHED AND SPLIT.

FIG. 198—THE PIECE OPENED OUT.

HOW TO MAKE A SLOT CIRCLE.

I will tell beginners how I make fifth wheels at the factory, I mean those with a slot in them—that is, how I make the slot part.

First make two pieces like that shown in Fig. 199, the parts *A A* being the same thickness as *B*, but twice as wide.

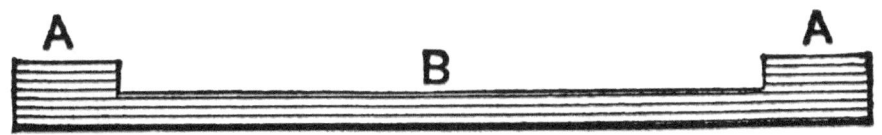

MAKING A SLOT CIRCLE BY THE METHOD OF "IRON DOCTOR."
FIG. 199—SHOWING THE SHAPE OF THE PIECE.

The distance between *A A* is made to suit the turning part of the job. Next weld the two pieces together at *A A* and form the slot *C*, as shown in Fig. 200, *DD* representing the *B* in Fig. 199, and *E* the ends welded together.

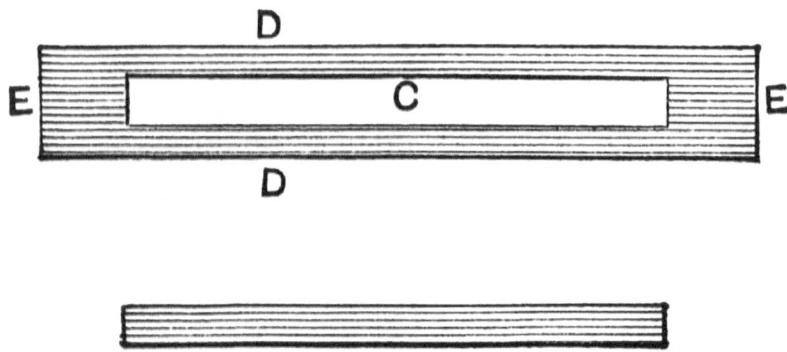

FIG. 200—SHOWING THE TWO PIECES WELDED TOGETHER, AND THE PIECE USED IN THE SLOT.

Then weld the slot of the clip and get the right lengths. Then fit a piece of iron of the shape shown in Fig. 201 into the slot so as to fill it up, then heat the whole to a red heat and bend around a former. Then the inside piece will contract. The filling is removed when the circle is complete.—*By* Iron Doctor.

FORGING A CLIP FIFTH WHEEL.

My way of forging a clip fifth wheel is as follows: Take a piece of Norway iron and fuller in a recess, as in Fig. 202, then punch in the two holes h, and split to the dotted lines, opening the piece out as in Fig. 203.

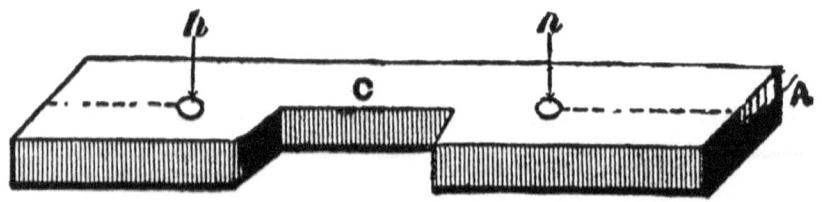

FORGING A CLIP FIFTH WHEEL BY THE METHOD OF "J. C. H." FIG. 202—SHOWING THE PIECE FULLERED AND PUNCHED.

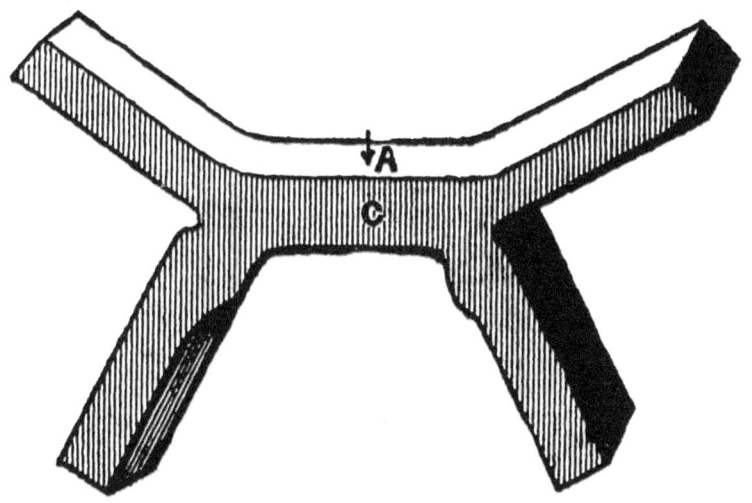

FIG. 203—SHOWING THE PIECE OPENED OUT.

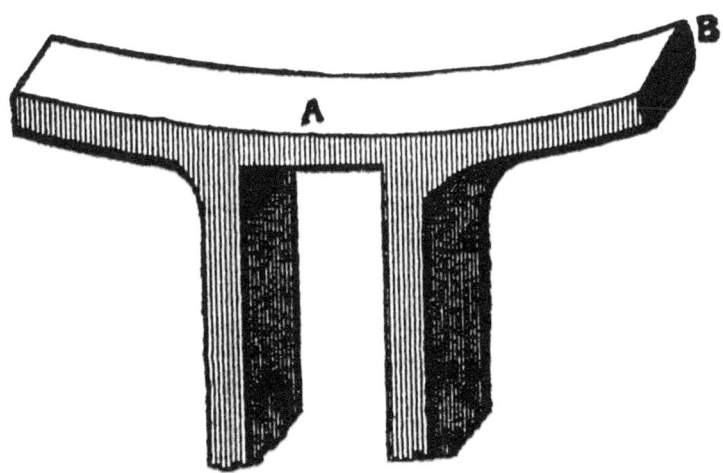

FIG. 204—SHOWING THE PIECE BENT TO SHAPE.

Next forge the whole to shape, draw out the shanks, cut the threads at their ends and then bend to shape, as in Fig. 204, the face *A* in Fig. 204 being the same face as *A* in Fig. 203. Weld at B (Fig. 204) the remainder to form the circle, or if the iron is long enough, and it is prepared, it can be made out of one piece.—*By* J. C. H.

METHOD OF MAKING FRONTS FOR FIFTH WHEELS.

With reference to the subject of fifth wheel making on a small scale, permit me to offer a wrinkle for the consideration of the craft.

The "bow" or "front" of many wheels used to be of a pattern represented by B in Fig. 205.

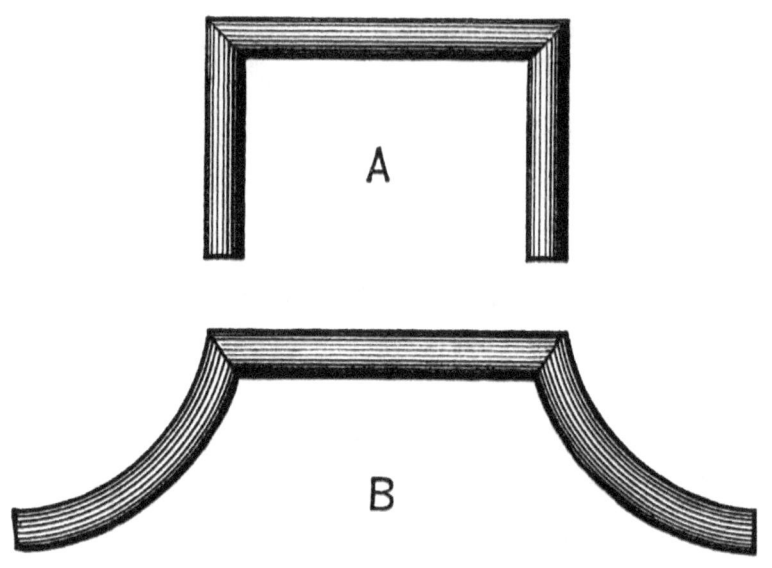

MAKING FRONTS FOR FIFTH WHEELS.
FIG. 205—SHOWING THE OLD METHOD OF MAKING THE MITRE.

It was made of ¼ or 5-16 iron. The usual way to form the mitres (which to look well require to be definite) was to upset the iron into a square angle at each mitre point, as at A, then open out and curve the ends to the desired form, and finish the mitres by considerable filing.

To facilitate this tedious plan (and it was tedious when thousands were required), I made the tools shown in Figs. 206, 207 and 208. A number of lengths—say 100 lbs.—of the rod, were cut about 5/8 inch for 5-16 inch and ½ inch for ¼ inch longer than the distance (measured with a piece of cord of the same diameter as the iron) between points *B* and *B*, Fig. 206, and these lengths were bent cold into the approximate shape or form shown by the dotted line B, Fig. 206.

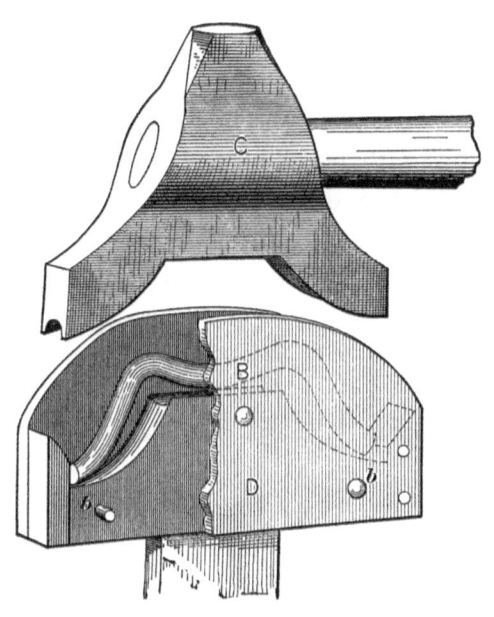

FIG. 206—SHOWING THE TOP TOOL AND SWAGE USED.

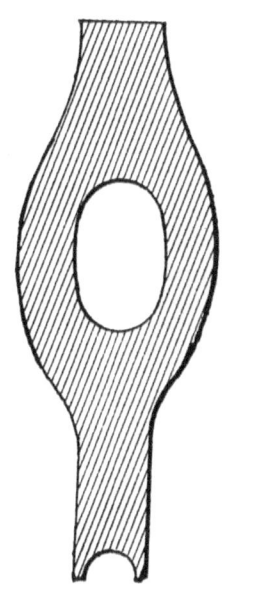 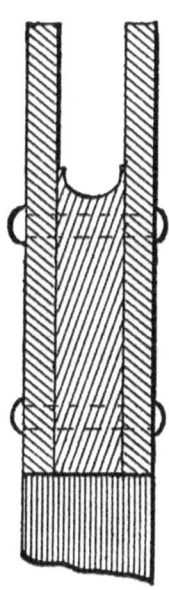

FIG. 207—END VIEW OF THE TOP TOOL.

FIG. 208—END VIEW OF THE SWAGE.

The swage and top tool being ready, a number of cold shaped blanks were put in the furnace or forge, and the helper taking out one at a time when white hot along the crown or middle, dropped it into the swage, when by two or three smart quick blows with the sledge upon the top tool *C*, Fig. 206, the bow or stay would be complete and ready for welding into the split ends of the circle top. The job I have described is completed without the use of a file. Done by the old plan, filing had been quite an item of expense and it is hardly necessary to say that I never went back to the old method.—*By* W. D.

MAKING A FIFTH-WHEEL HOOK OR A POLE STOP.

To do any work properly and quickly we require the right kind of tools. In making a fifth-wheel hook on a pole stop we need a heading tool, as shown in Fig. 209. *A* is the thick heading portion.

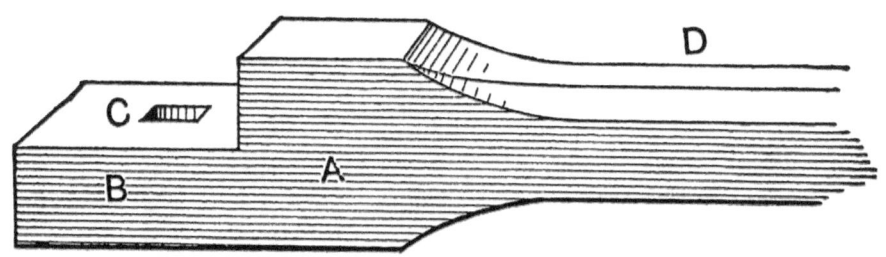

MAKING A FIFTH-WHEEL HOOK OR A POLE STOP.
FIG. 209—THE TOOL USED.

B is the thin heading post. *C* is the square hole. *D* is a section of the hard portion. Take good iron of the proper size, draw down and form the end *D*, shown in Fig. 210, to fit the square hole *C* in the tool. Heat the iron, insert in the tool, and split with the splitting chisel, as at *E*. Next, with a small fuller open up the split, as at *K*, Fig. 211, sufficiently to insert a larger fuller without continuing the split further down. The part *F* is inserted in the tool, and *H* is thrown off. Next take a larger fuller and force down *H*, as at *N* in Fig. 212, *L* fitting in the tool while *M* is still upright. Twist *N* to the shape desired, then repeat and insert the part *O*, Fig. 213, in the tool, hold the sledge on *S*, turn down *P* and dress up the whole with the set-hammer.

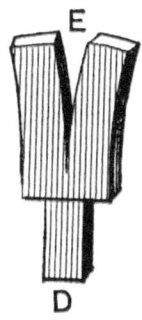 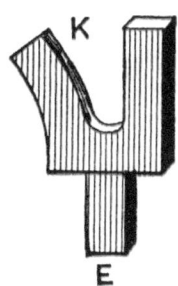

FIGS. 210 AND 211—SHOWING HOW THE PIECE IS SPLIT AND FULLERED.

R is the space or recess covering the fifth wheel when the device is used as a fifth-wheel hook, and resting on the cross-plate when it is used as a pole stop.

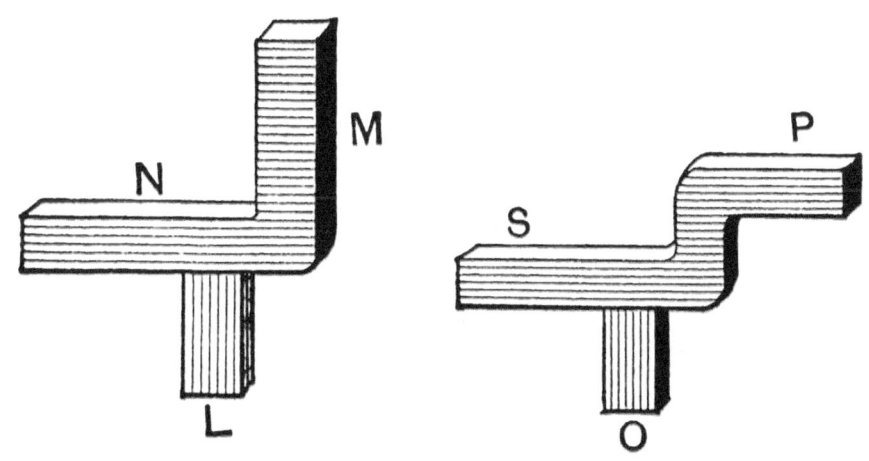

FIG. 212—SHOWING THE NEXT STAGE IN THE FULLERING PROCESS.

FIG. 213—SHOWING THE JOB COMPLETED.

Remove the piece from the tool and round up the end *O* to the proper size for the thread and nut. The square portion is for a fifth-wheel hook, and should be left just as long as the iron is thick through which it passes, less enough to give a set or tension with the nut. The thread should be cut clear up to the square portion. The nut ought to be over the standard thickness to admit of having an extra thread, with a view to getting a better grip.

For a pole stop make the shank the same as you would a bolt. The beginner may require four or five heats to make one. A little practice will reduce the labor to three heats. We have seen experts do them in two heats.—*By* Iron Doctor.

MAKING A SHIFT RAIL.

The accompanying sketches represent a new and inexpensive method of making a shift rail for buggy seats, as the following explanation will show:

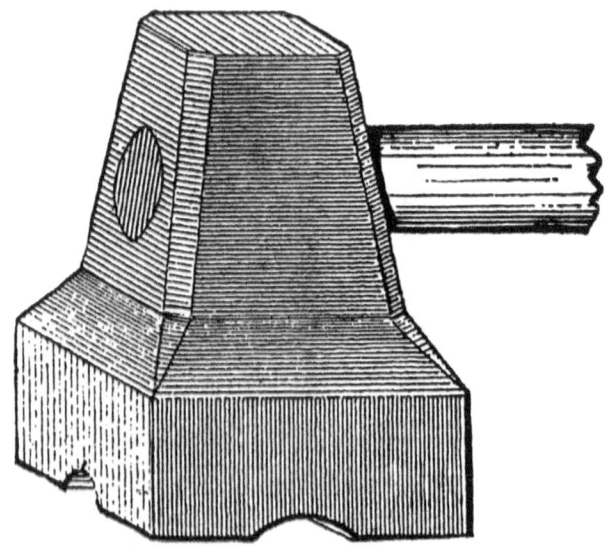

MAKING A SHIFT RAIL.
FIG. 214—UPPER PART OF THE SWAGE.

Take a piece of 5/8 by 7-16 inch oval Norway iron, upset it at A, in Fig. 216, then take an oval-pointed cold chisel and make a hole in the place where it was upset large enough to admit a piece of ⅜-inch round iron, with a collar swaged on one end of it as shown in Fig. 216.

Take a good welding heat at the point where the two irons intersect, and then place in the swage shown in Fig. 215. Place the swage shown in Fig. 214 on top and strike a few blows with a sledge hammer.

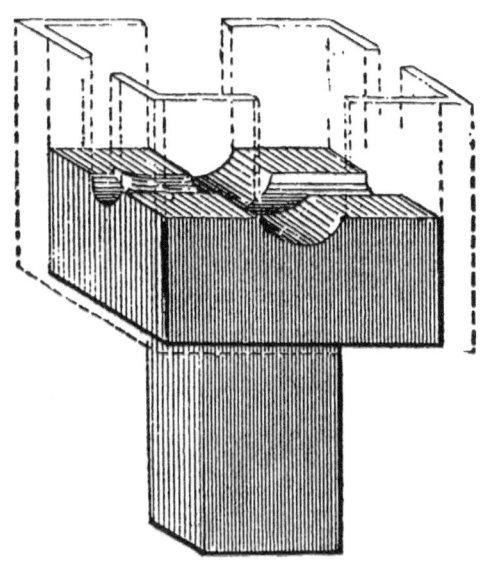

FIG. 215—LOWER PART OF THE SWAGE USED IN SHAPING THE RAIL.

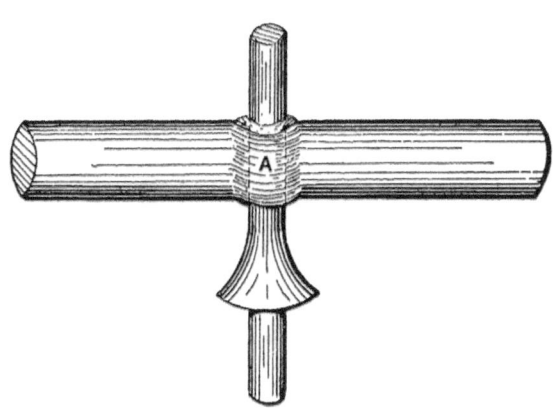

FIG. 216—THE SHIFT RAIL COMMENCED.

The result will be a forging of the kind shown in Fig. 217, ready for welding on the goose neck and arm rests, as shown by the dotted lines. In strength this construction is about equal to a forging made out of solid iron, while it is exactly the same in looks. The swages shown in Figs. 214 and 215 are made the same as other swages employed by blacksmiths, except that those shown in Fig. 215 should be provided with a piece of band-iron arranged as indicated by the dotted lines.

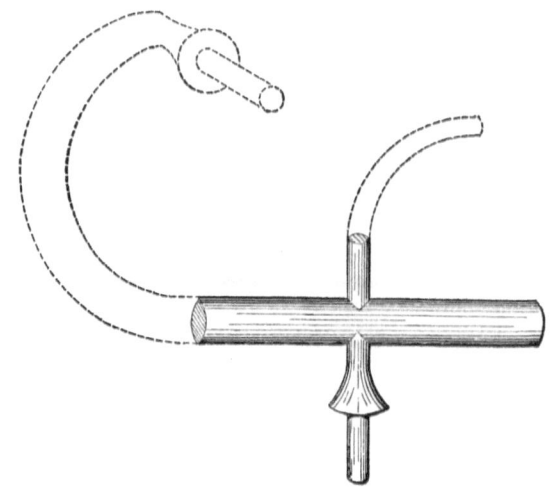

FIG. 217—THE FINISHED FORGING READY FOR WELDING.

This should be shrunk around it. It projects above the face of the swage a distance of three-quarters of an inch or more, and serves the purpose of holding the top swage in correct position while taking the weld.—*By* C. T. S.

MAKING SHIFTING RAIL PROP IRONS BY HAND.

The following description of some tools for making shifting rail prop irons by hand, with sketches showing how to make an iron in the tools and how to prepare it for the tools, describes my method of making shifting rail prop irons.

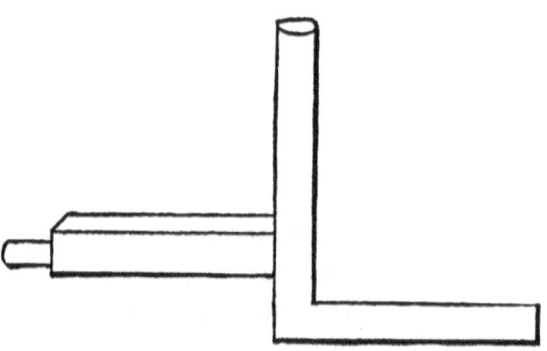

FIG. 218—SHOWING THE FINISHED PROP IRON.

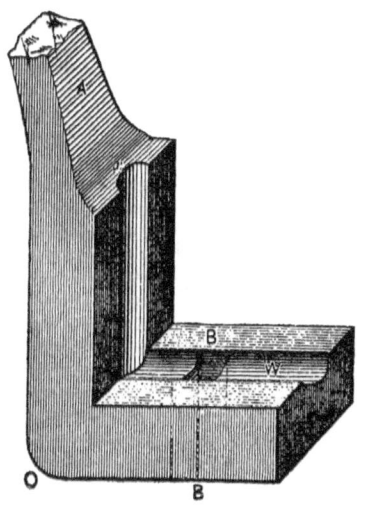 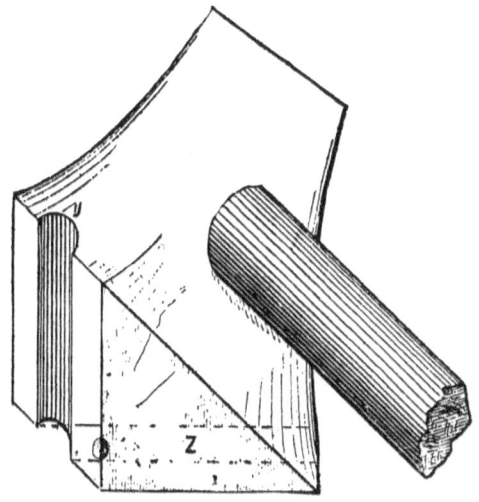

MAKING SHIFTING RAIL PROP IRONS. FIG. 219—SHOWING THE TOOL USED TO GIVE THE IRON A COMPLETE FINISH.

FIG. 220—SHOWING THE TOP SWAGE.

Fig. 218 shows the prop iron as it is when finished. To make this take the proper size of iron and split it out as in the dotted lines of Fig. 222, which will give two irons as in Fig. 223.

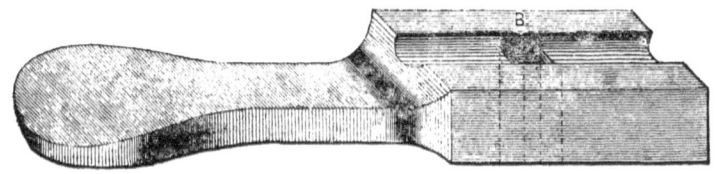

FIG. 221—SHOWING THE TOOL TO WHICH THE SQUARE A IS FITTED.

With the same heat used in splitting out the two pieces draw the square A, shown in Fig. 223, so it will fit the square B of the tool shown in Fig. 221; now split Fig. 223 as at X, and with the same heat place the square part A in the square part of the tool, Fig. 221, at *B*; then turn down the part *O*, then, with an ordinary top swage, swage it down; now take the iron out of the tool and reheat it, edge it up and place it in the tool Fig. 219, and use the top swage, Fig. 220.

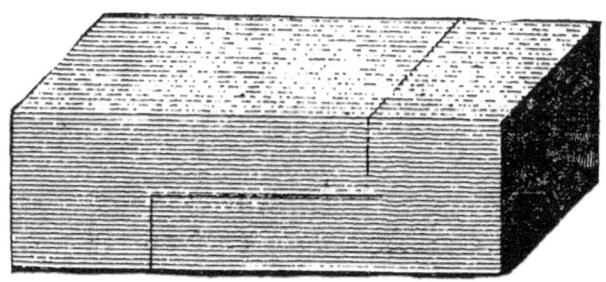

FIG. 222—SHOWING THE LINES ON WHICH THE IRON IS SPLIT

The tool, Fig. 227, gives the iron a complete finish and makes all irons alike at the corners, which is a big gain where irons are to be duplicated. The tool, Fig. 219, is used on the anvil the same as any ordinary heading tool.

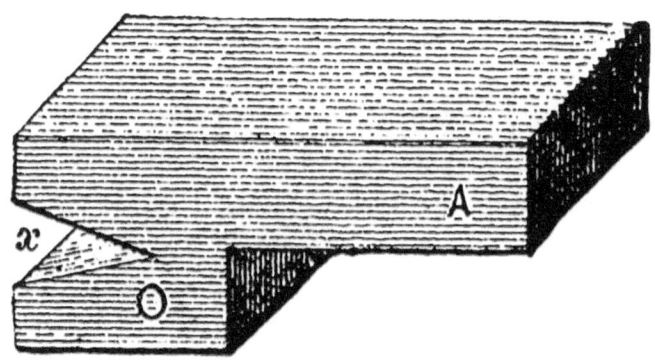

FIG. 223—SHOWING THE IRON SPLIT.

Hold the corner on the anvil while swaging the corners; while swaging over the square part set the square part of the prop iron in the square hole of the anvil, and so on. In a short time any ordinary smith will be able to handle the tools properly and make a good iron. I do not think it necessary to give any further description of the tools, as I think the sketches will speak for themselves.—*By* H. R. H.

GETTING OUT A SOLID KING-BOLT SOCKET.

My way of making a solid king-bolt socket is as follows:

I first make a block as shown in Fig. 226. It is about four and one-half inches long, and if the plate is to be one and one-quarter inches wide, make the block, A, two and one-half inches wide and one and three-quarter inches thick; make the swage part, B, one and one-quarter inches half round; make the hole nearly one and five-sixteenths inches in diameter at the top and taper to one and one-quarter inches on the bottom, which permits of the easy removal of the iron.

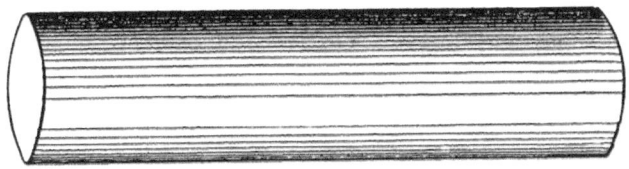

GETTING OUT A SOLID KING-BOLT SOCKET BY THE METHOD OF "IRON DOCTOR." FIG. 224—SHOWING THE IRON BEFORE SPLITTING.

Then take round iron one and one-quarter inches in diameter and five inches or more long, as shown in Fig. 224, split, as shown in Fig. 225, three inches down, and open up; split with a dull chisel or round fuller, swage the joint of round portion just a trifle so as to allow it to enter the block easily, reheat to a moderate heat and then insert the tool in the block. If the square hole in the anvil is not large enough, or a swage-block is not at hand, hold the piece over the open forging vise and bend the ends down moderately until nearly a right angle is formed.

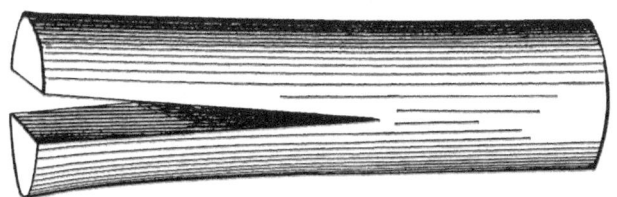

FIG. 225—THE IRON AFTER SPLITTING.

Next cut off the end so that it will be less in length than the block is thick by a matter of three-eighths of an inch, raise a welding heat, insert the point in the block and flatten or set down into the swage—first with hammer and sledge, striking quick blows so as to weld up any looseness, then use the flat hammer to finish—as a result we get Fig. 227—in which *A* is the plate and *B* the stem or portion to be converted into a socket. Then weld in iron of the proper size, draw and swage to a finish and fit while the iron is cooling off.

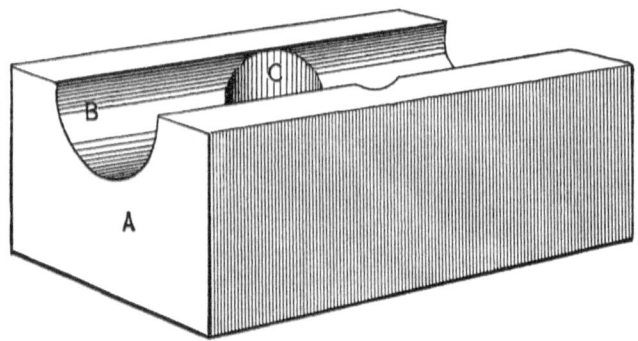

FIG. 226—THE BLOCK.

I will say just here that it is very improper to fit iron to wood by burning it on as some stupid blacksmiths do.

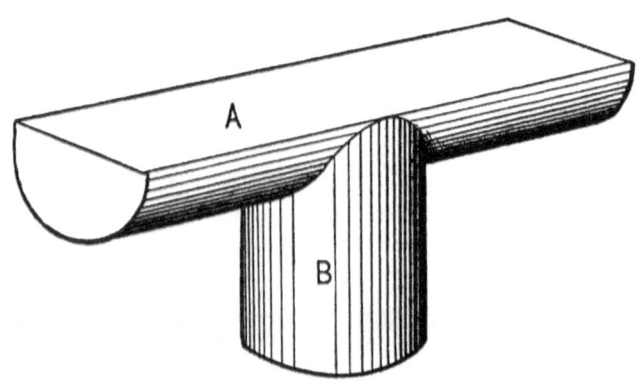

FIG. 227—THE PLATE AND STEM.

Hit the iron while it is black hot, and should it still burn, rub chalk over the portion which comes in contact with the iron.

The socket is finished inside with two drills, one for the insertion of the smaller portion of the socket, the other to make a passage for the king-bolt, which in this case may be five-eighths of an inch.

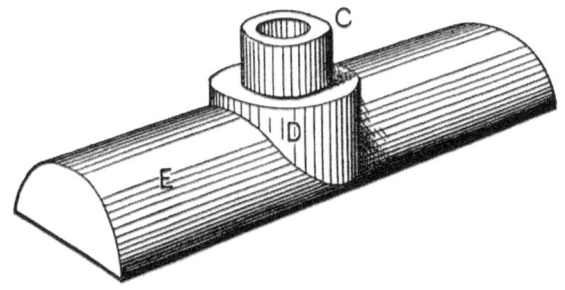

FIG. 228—THE PLATE AND SOCKET.

In Fig. 227, *A* denotes the plates; *B* the full-sized part of the socket and the same diameter as the socket; *C*, of Fig. 228, is the small portion of the socket which fits in the socket and is formed by swaging.

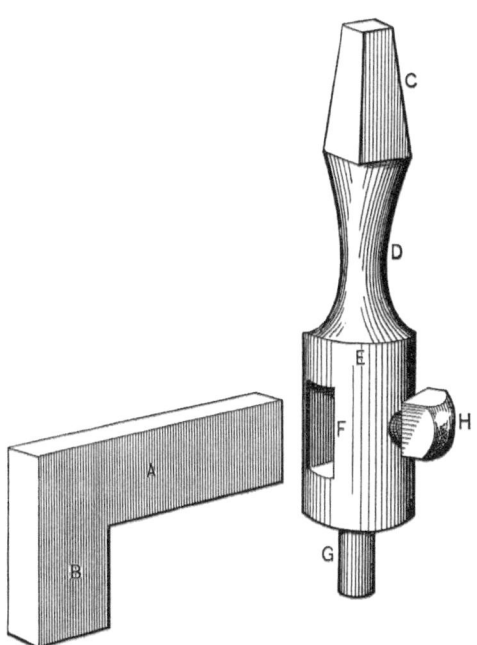

FIG. 229—SHOWING A TOOL FOR TURNING OFF THE UPPER PORTION OF THE SOCKET.

Fig. 229 represents a good tool for turning off the upper portion of the socket; A is the portion which fits in the drill-chuck; B the waist, about three-quarters of an inch diameter; two inches long; C, the body of the tool, one and one-quarter inches diameter, two inches long; hole F may be one inch by one and one-half inches for the insertion of the portion marked A, about three and one-half inches long, one inch by one and one-half inches. B is the cutting part. In Fig. 229 is shown a set-screw to hold the cutter fast. The point E fits the hole in the socket and acts as a guide. With a little practice the tools shown in Fig. 228 will answer as well as a lathe.—*By* Iron Doctor.

HEADING BOLTS.

I send a sketch in Fig. 230 showing a heading block for an anvil, the mouth of the hole being worn rounded, as shown at *b b*, which would let the iron fill in, as at *c c* in Fig. 231. Now a very little wear there will let the iron fill in enough to make quite a difference in the length of blank required to make a bolt-head to any given dimension.

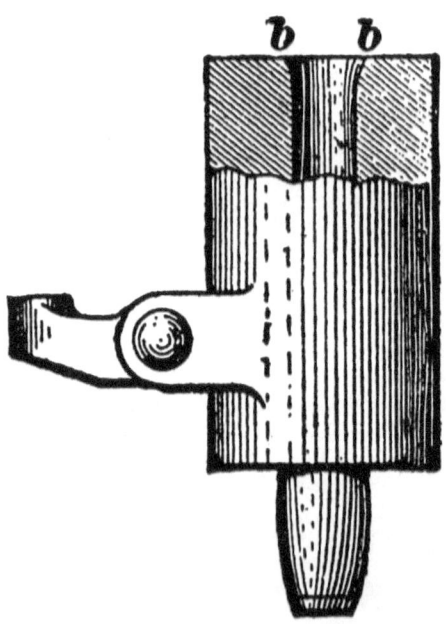

FIG. 230—SHOWING THE HEADING BLOCK FOR THE ANVIL.

But the extra iron is not of so much consequence as the swell in the bolt neck, which is a great nuisance, especially where both are to be threaded clear up to the head. In the anvil block used in our shop we have steel dies set in the block, as in Fig. 232 at *d*. By this means different sizes or dies may be fitted to one block.

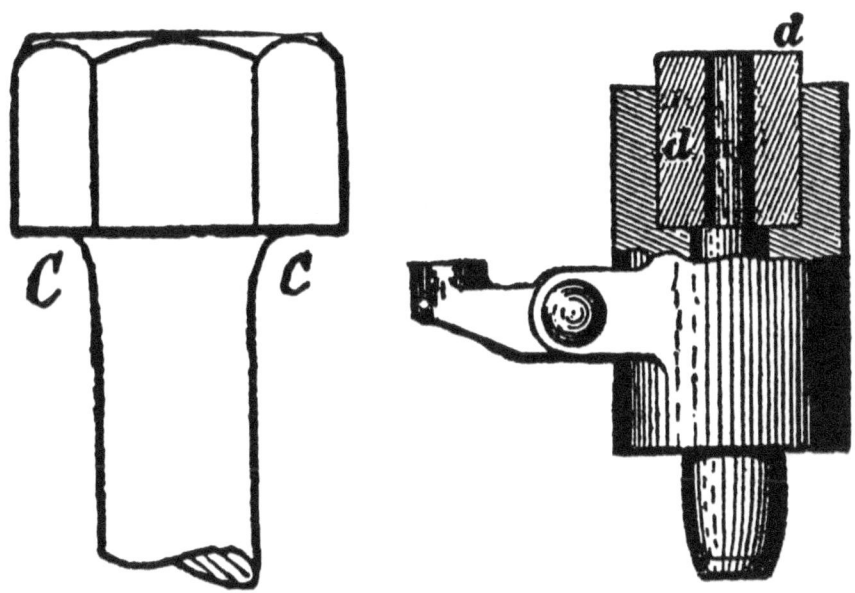

FIG. 231—SHOWING THE IRON FILLED IN AT C C.

FIG. 232—SHOWING THE STEEL DIE SET IN THE ANVIL BLOCK.

The dies may be turned upside down, by making the dies d longer than the recess is deep, so as to stand above the top of the block. We are enabled to heat the dies and close up the holes when they have become too much worn.—*By* J. R.

HEADING BOLT BLANKS.

In heading hexagon bolts I allow three times the diameter of the iron, providing the heading tool is the proper size. Suppose, for example, a ¾-inch bolt is to have a stem three inches long.

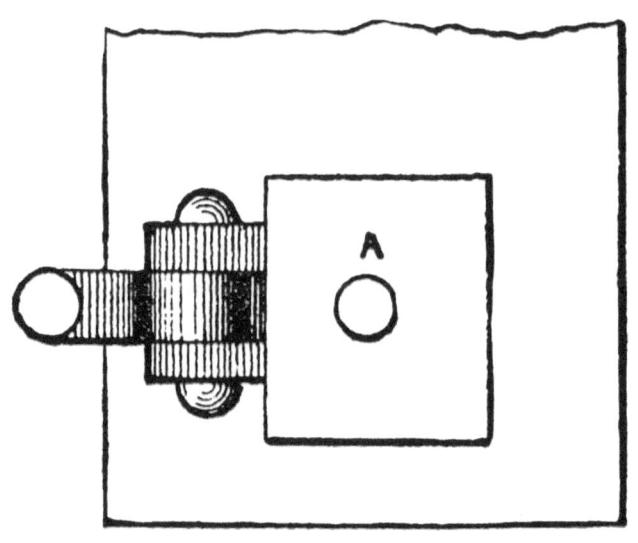

A SUBSTITUTE FOR A BOLT-HEADER, AS DESIGNED BY "J. R." FIG. 233—TOP VIEW.

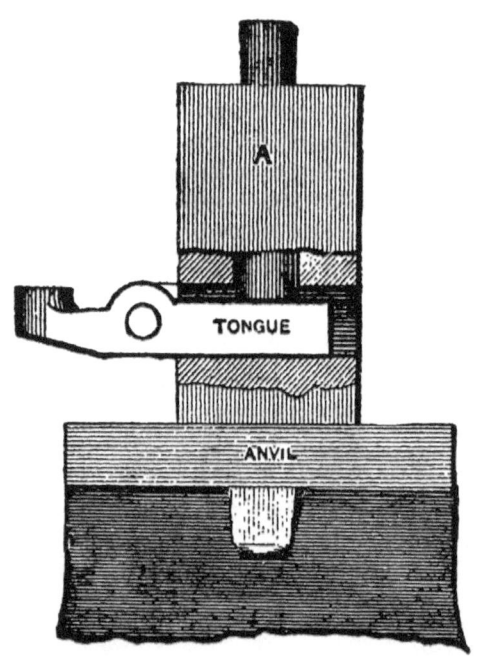

FIG. 234—END VIEW.

Then in cutting the iron rod into blanks I make the blanks five and one-fourth inches long, which allows two and one-fourth inches of blank length to form the head, the bore of the heading tool being ¾ and 1-64th. The header, or the dies, as the case may be, should by rights be of steel, as cast-iron wears very quick; and after five hundred bolts have been made in a cast-iron die there will be so much wear that the heads will not come up to size with the above allowance.

A good shop tool for short bolts, which I like better than a bolt-header, consists of a block *A*, Fig. 233, with stem to fit in the square hole in the anvil, and having a slot through it to receive the usual tongue.

Fig. 233 is a top view of the tool, and Fig. 234 is an end view.—*By* J. R.

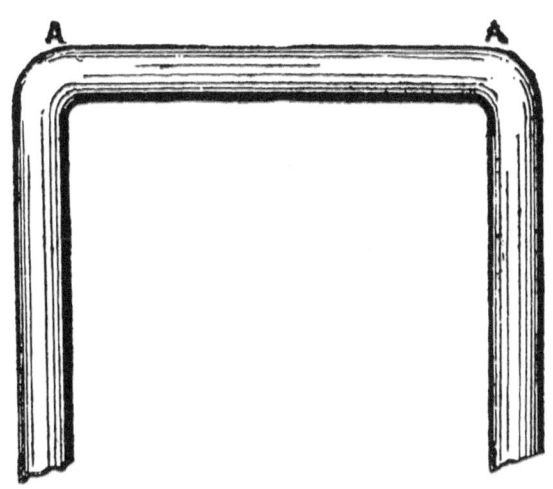

BENDING A CAST-STEEL CRANK.
FIG. 235—SHOWING THE METHOD OF BENDING AT A A.

BENDING A CAST-STEEL CRANK SHAFT FOR A TEN HORSE-POWER ENGINE.

The illustrations herewith will show how I made a crank, the shaft for which was two and one-half inches in diameter. I made the bends at *A* and *A* first, and shaped as shown in Fig. 235. I then heated at *B*, cooled off *A*, and bent as seen in Fig. 236.

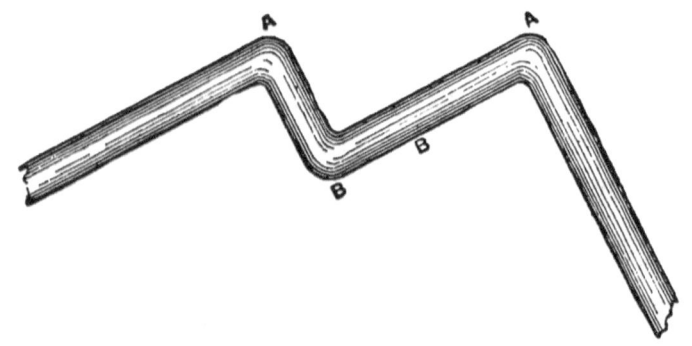

FIG. 236—SHOWING THE CRANK BENT AT A, A AND B.

I then heated at *A, B* and *B,* cooled off *A*, and bent as shown in Fig. 237. From *A* to *A* in Fig. 235 I allowed fourteen inches to make a five-inch stroke.

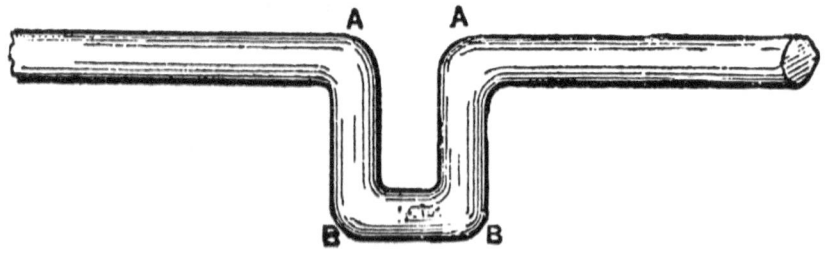

FIG. 237—SHOWING THE JOB AS COMPLETED.

I straightened so as to turn to two and three-eighths inches, and there was not a cockle or a scale on it.—*By* Southern Blacksmith.

MAKING A CLEVIS.

A man can learn his trade with a good mechanic, and then go in another shop and see the same piece of work done in another way, much simpler and easier than his. To illustrate this point, I will explain what I once saw a good workman do. The job I refer to was the making of a clevis. He wished to make a five or six-inch one as follows: Large in the center, tapering forward to the bolt holes, the holes to be flat on the inside and round on the outside.

MAKING A CLEVIS BY THE METHOD OF "T. G. W." FIG. 238—
SHOWING THE ORDINARY WAY.

That required the iron to be half round to form the collar for the bolts. To do this he cut the iron to the length required, drew out the ends, bent it over the horn of the anvil, then over a mandrel, took a heat, welded down, and then hammered it over the horn again. By this time it required another heat, as it needed more drawing, and this way of doing the job took twice the time really necessary for it.

FIG. 239—SHOWING THE PIECE AS DRAWN OUT AND FLATTENED.

He then finished the other end in the same way, bent it, and this completed the work, the collar being then as shown in Fig. 238 of the accompanying illustrations.

Now, he could have saved both time, coal and labor by having taken a piece of iron the required size, drawn out the ends, then laid it in a swage and taken his set; then flattened it back two and one-half inches, or one and one-half inches, from the ends, flattening say three inches in the swage.

FIG. 240—SHOWING THE CLEVIS AFTER BEVELING AND WELDING.

This would have made it as shown in Fig. 239; and after beveling and welding the collar the holes would have been as shown in Fig. 240. —*By* T. G. W.

CRANK SHAFTS FOR PORTABLE ENGINES.

Herewith I send you sketches of my plan for making crank shafts for small portable engines.

Fig. 241 shows the pieces composing the crank, and the other figures the details, having on them the same figures of reference.

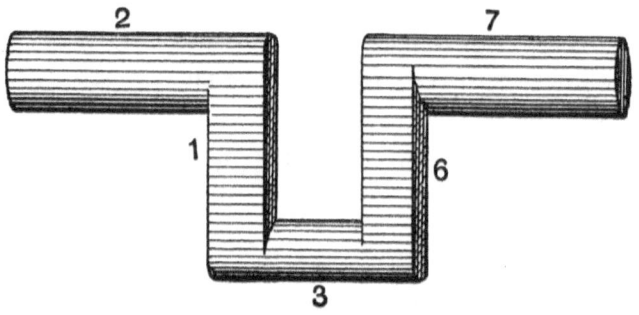

FIG. 241.

FIG. 242.

First make two pieces, as shown at 1 and 6 in Fig. 241, and weld Fig. 242 to piece 1; then weld Fig. 243 to piece 1; next weld around the ends of 1 in Fig. 244 two straps, these straps being shown at 4 and 5.

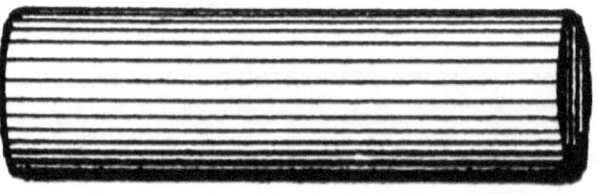

FIG. 243.

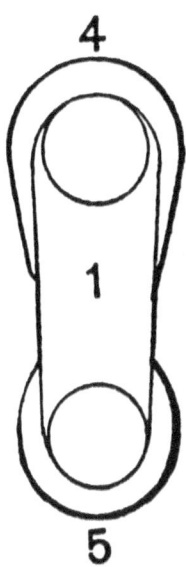

FIG. 244.

I weld up the other half in like manner. I sometimes make cranks in this way that weigh up to 150 lbs.—*By* Southern Blacksmith.

FORGING A LOCOMOTIVE VALVE YOKE.

Plan 1.

My method of making a valve yoke and stem for locomotive engines requiring but one weld (these parts are usually made with three welds; the saving accordingly will at once be perceived) is as follows: I take a car axle, or any iron of the required size, heat and weld it properly, set it down with a fuller to form the corners, allow for draw between each corner, and then draw to the required size as shown in Fig. 245. I leave plenty of stock in the corners so it can be trimmed with a chisel as necessary. I next bend as shown in Fig. 246. To make the stem I use two-inch square iron, which I twist and weld up the entire length to change the grain of the iron, so that it will not cut the packing or stem. I leave the end for welding on to the yoke as shown in Fig. 246. I take separate soft heats on each, being sure to have them clean, and then weld.

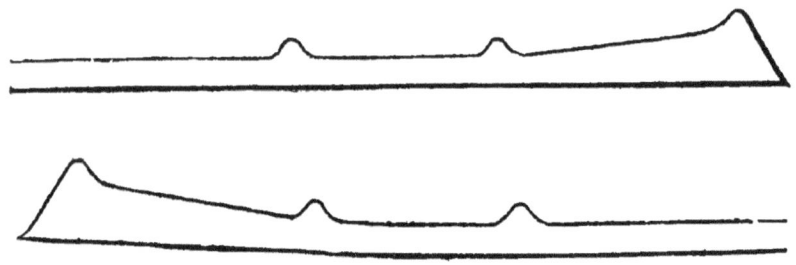

LOCOMOTIVE VALVE YOKE.
FIG. 245—THE PIECE BEFORE BENDING.

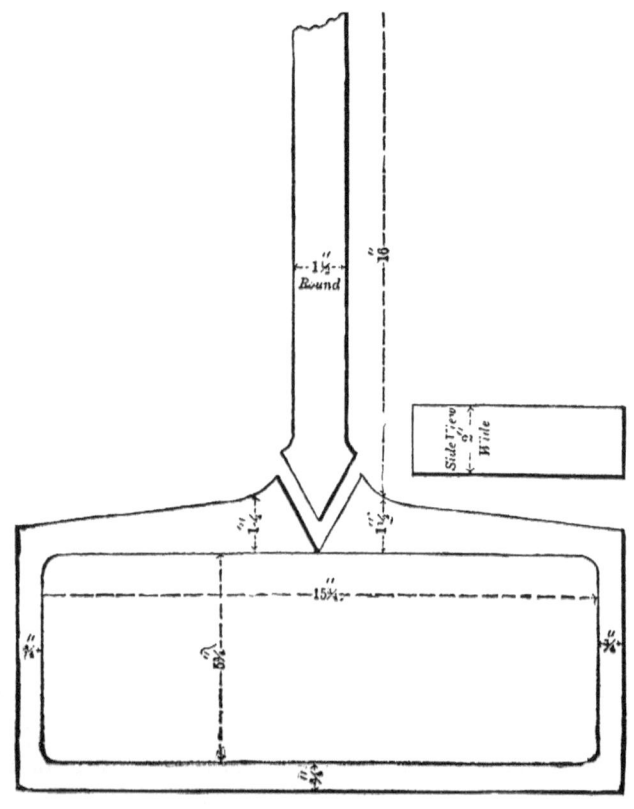

LOCOMOTIVE VALVE YOKE.
FIG 246—FORMED, READY FOR THE FINAL WELD.

Although I use no clamps, I have never had any trouble from the valve yokes springing. I allow that side to come a little long, as it is easier to stave work of this kind than it is to draw it. My reason for considering a V weld the best is that we have more surface to weld in a V weld than we could get in a jump weld; besides, we get into the *body* of the work that is to be done. Any blacksmith who will subject the two welds to a fair test will soon be convinced that a jump weld will show flaws and break under less pressure than a V weld. Why do most smiths V steel in bars, sledges and other similar work if lap or jump welds are best?

The usual way of making articles of this kind is to have a weld in each forward corner or in each side, having a jumped weld for the stem. By the plan that I pursue a V weld is used, which is the best that can be made. The oil and steam will soon destroy a jumped weld, as they are frequently unsound. The sketch is reduced from full size. The figures, however, denote the dimensions of the finished work.—*By* R. O. S.

FORGING A LOCOMOTIVE VALVE YOKE.

Plan 2.

I think that the plan I show in the sketch herewith is a good way to forge a locomotive valve yoke, and it has the advantage of having only one weld.

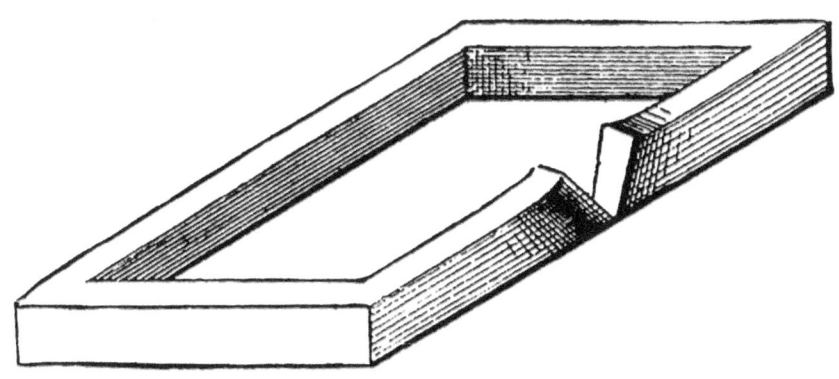

FORGING A LOCOMOTIVE VALVE YOKE, "UNKNOWN'S" METHOD.
FIG. 247—YOKE BENT AND SCARFED READY FOR WELD.

FORGING A LOCOMOTIVE VALVE YOKE.
FIG. 248—ROD SCARFED FOR WELD.

FORGING A LOCOMOTIVE VALVE YOKE. FIG. 249—ROD IN POSITION FOR WELDING.

The yoke is forged around as in Fig. 247, and scarfed up for the rod weld, which is formed as in Fig. 248, the weld being made as in Fig. 249.—Unknown.

FORGING A LOCOMOTIVE VALVE YOKE.

Plan 3.

The following plan of making a locomotive valve yoke is followed in my shop: We cut an axle to the length desired and draw the stem first, leaving it square as shown in Fig. 250. We then draw the other end, and punch and cut under the hammer until we get the shape shown in Fig. 251.

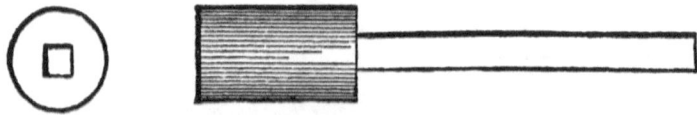

MAKING A VALVE YOKE BY THE METHOD OF "NOVIS HOMO."
FIG. 250—SIDE AND END VIEWS OF THE STEM AFTER DRAWING.

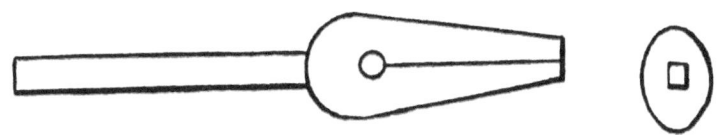

FIG. 251—SIDE AND END VIEWS OF THE OTHER END AFTER DRAWING, PUNCHING AND CUTTING.

We next split, straighten and open out as seen in Fig. 252. The ends are then drawn out on a table such as is shown in Fig. 253. The stem is then twisted by placing the opened ends under the hammer, and applying to the square end the lever shown in Fig. 254.

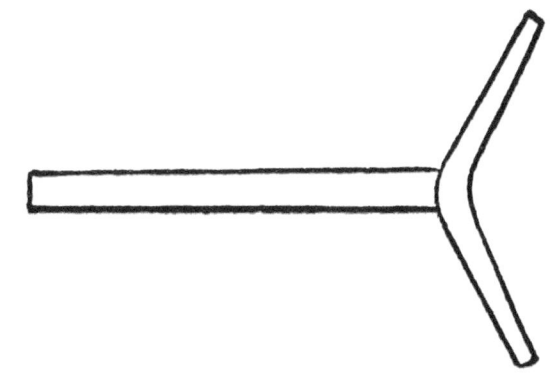

FIG. 252—SHOWING THE PIECE OPENED.

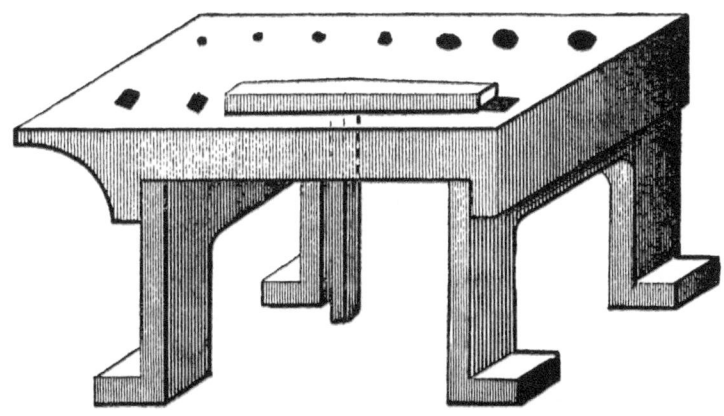

FIG. 253—SHOWING THE PIECE STRAIGHTENED IN THE TABLE.

FIG. 254—SHOWING THE LEVER USED TO TWIST THE ROD END.

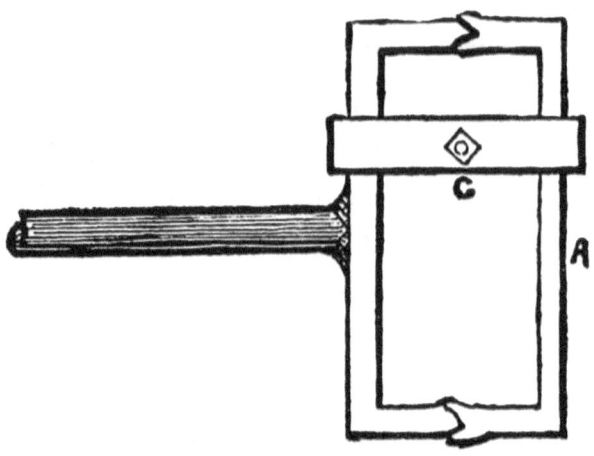

FIG, 255—SHOWING THE PIECE C DRAWN FOR WELDING.

A piece of iron like that shown at *A* in Fig. 255 is then forged down, the two valves of the yoke are put together, and we weld as shown in Fig. 255, bolting the two pieces together by the clamp *C*. Or we forge the second piece longer on its ends and weld as in Fig. 256.

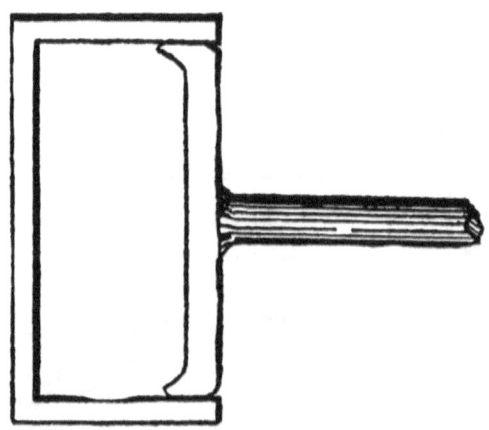

FIG. 256—SHOWING ANOTHER METHOD OF WELDING.

We have also a face plate designed especially for making valve yokes. It is like the one shown in Fig. 253, excepting that it has but one hole. In speaking of hammers I mean steam hammers.—*By* Novis Homo.

FORGING A LOCOMOTIVE VALVE YOKE.

Plan 4.

Locomotive blacksmithing is one of the most important branches of the trade, and requires all the skill and experience that the smith can command.

LOCOMOTIVE VALVE YOKE.
FIG. 257—SHOWING YOKE JUMPED AND DRAWN FOR BENDING.

I do not know that I can show a better way than the preceding, but I will attempt to describe my way, which I think a good one. In the first place, after having got the required size iron, I jump the stem and draw both ends (all as shown in Fig. 257).

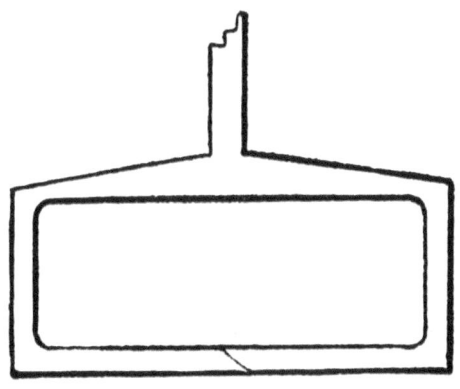

FIG. 258—SHOWING YOKE BENT AND WELDED.

The next operation is simply to bend into shape and weld as shown in Fig. 258.—*By* Vulcan.

FORGING A LOCOMOTIVE VALVE YOKE.

Plan 5.

Many yokes lack durability because the oil and tallow used in oiling the valves contain some substance which eats into the fiber of iron if it can get at the end of the bar. I have seen many yokes made in "Unknown's" manner that were eaten in from one-half to one and one-half inches at the end of the stem. If in welding steel on the stem close to the yoke—that is, the stuffing box—you do not get a perfect weld, or if a flaw is left at the point of the steel or the scarfs of the iron, the corrosive substance to which I have alluded will soon eat a hole in the stem, and perhaps all through it.—*By* R. T. K.

FORGING A LOCOMOTIVE VALVE YOKE.

Plan 6.

While I am, as a rule, in favor of the V weld, I do not like its application to the valve yoke, as I believe it is liable to spring from being so weak in the ends and the opposite side.

I prefer a yoke made on "Unknown's" plan, because that obviates the danger of "springing" while the stem is being welded.

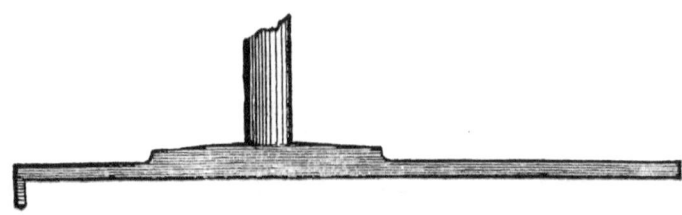

MAKING A LOCOMOTIVE VALVE YOKE BY THE METHOD OF "R. D."
FIG. 259—SHOWING THE PIECE READY FOR BENDING.

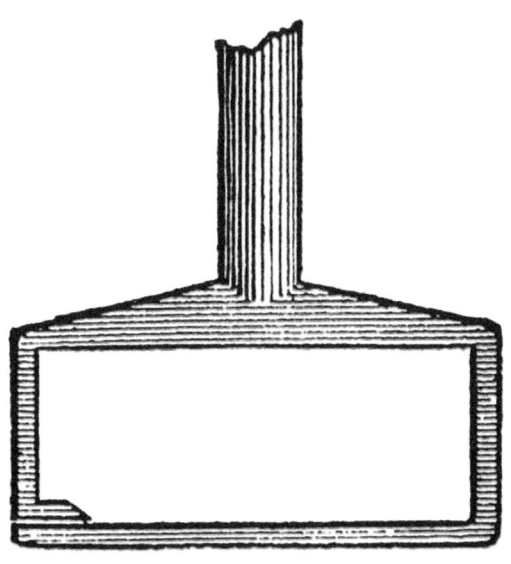

FIG. 260—THE YOKE BENT.

My own method is first to bring to the shape represented in Fig. 259 of the accompanying engravings. The yoke I make has the two ends and side of the Same size. I then bend as shown in Fig. 260. By this plan I save the time needed to make a corner.—*By* R. D.

FORGING A LOCOMOTIVE VALVE YOKE.

Plan 7.

Before giving my method of forging valve yokes I shall criticise briefly "R. O. S.'s" method of making them. He finishes the yoke where I should begin it.

FORGING A VALVE YOKE BY THE METHOD OF "R. O. S."
FIG. 261—SHAPE OF IRON AT START.

He says: "I first get out my iron in the shape shown in Fig. 261 (see the accompanying illustrations). I then bend as in Fig. 262." It will be seen that "R. O. S.," in his method, has four corners to form of the right shape and length before welding on the shank A; consequently, if he is unlucky—and this may happen to the best of workmen—his labor is all lost, and the iron goes to the scrap pile.

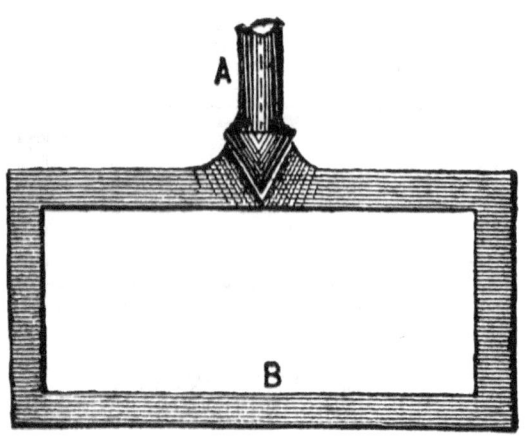

FIG. 262—THE PIECE BENT TO THE YOKE FORM

No one can weld A to B without stretching the yoke, and when this occurs the yoke assumes the shape shown in Fig. 263, thereby making it necessary to go all over the work again.

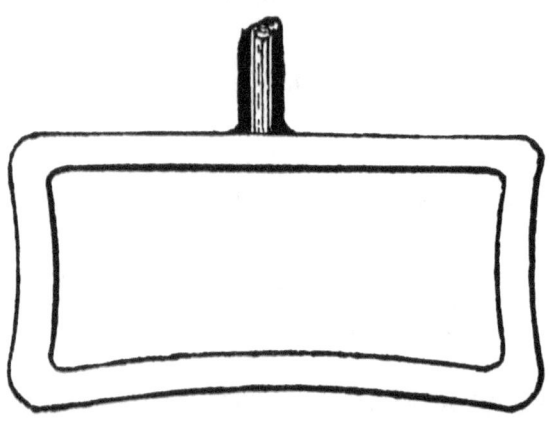

FIG. O. S."

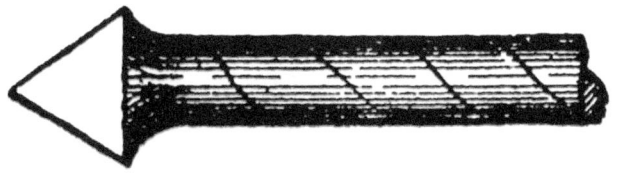

FIG. 264—SHOWING THE STEM TWISTED.

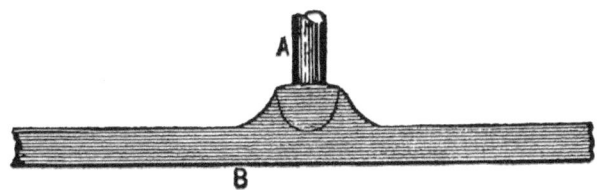

FIG. 265—THE IRON DRAWN INTO SHAPE.

Then "R. O. S." twists the stem *A* to change the grain of the iron, so that the packing on the stem will not be cut out. In twisting and welding up for the stem as he suggests, you twist with the iron whatever is in it, and in so doing you make a spiral of the stem as shown in Fig. 264, and if there is any hard substance left in the iron it is scraped out through the packing by the motion of the rod, somewhat in the manner in which a rat-tail file would operate in drawing it back and forth in a hole.

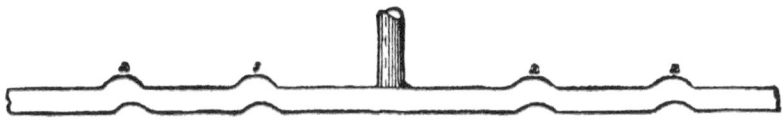

FIG. 266—THE SECOND STAGE IN THE JOB.

If you do not twist your rod, although it has sand in it, you get your cut in one or two places, according to the purity of the iron. Twisting does not remove impurities, neither does welding, so the time consumed in such work is lost, for bar iron as clear as steel can be got for what this twisting and welding would cost.

Perhaps some of the plans of my own, which I shall now give, may not seem feasible to all of the craft, but I have found every one of them satisfactory. I have six plans. The first is as follows: I take a piece of good iron, draw it out into shape as shown in Fig. 265, then form as shown in Fig. 266, and weld at C, by splitting as in Fig. 267.

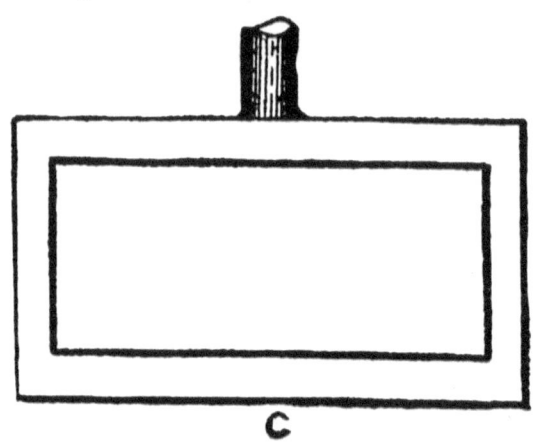

FIG. 267—THE YOKE COMPLETED.

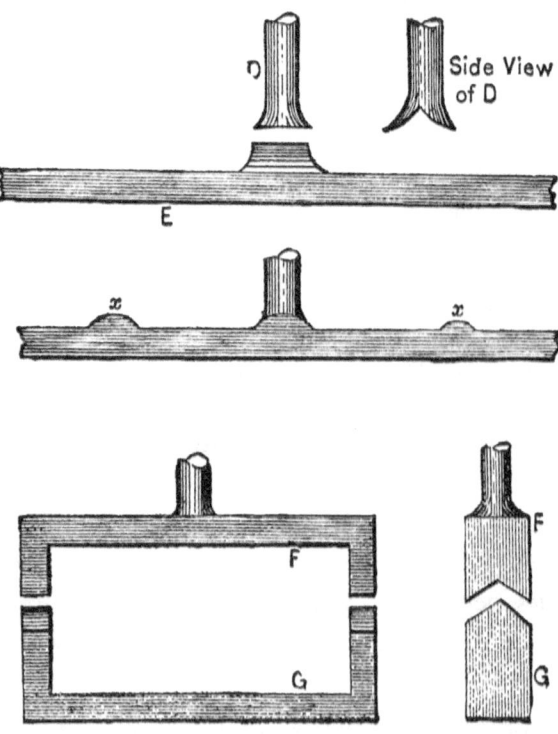

FORGING A VALVE YOKE, BY "J. T. B."
FIG. 268—THE SECOND METHOD.

Fig. 268 shows my second way of forging a valve yoke. I jump *D* to *E*, by splitting *D*, then forge, leaving projections at *X* to bring up the corners, shape the bend around, as at *P G*, split *F*, forge a piece *G* and weld *P* and *G* together.

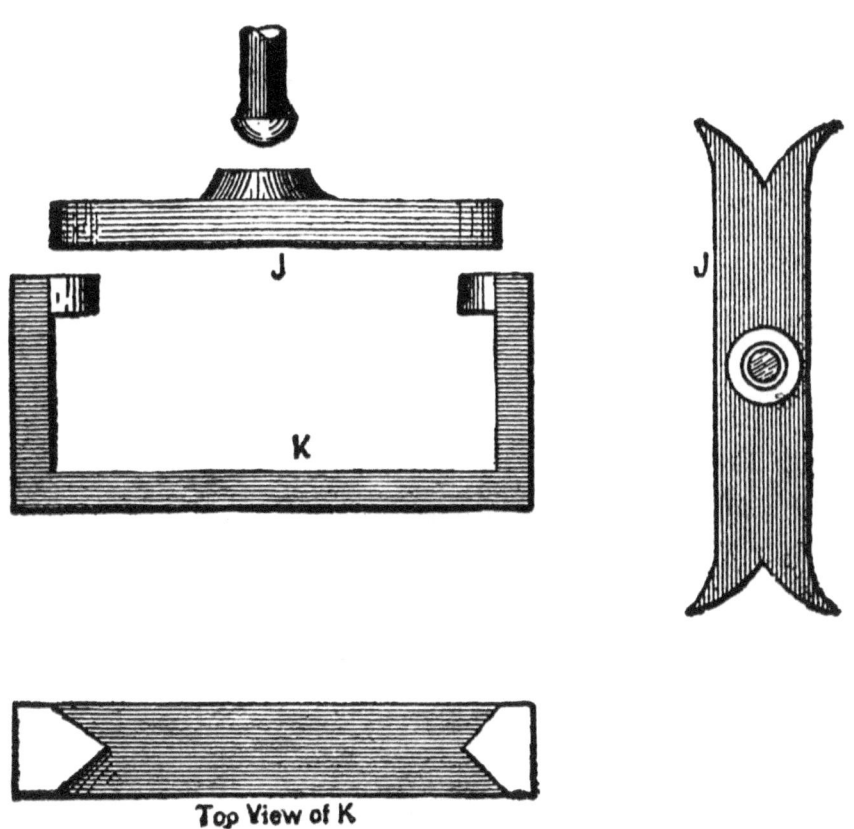

FIG. 269—THE THIRD METHOD

In my third method, shown in Fig. 269, I round the end of the stem, and open the ends of the part *J*, then forge a separate piece *K*, and weld *J* and *K* together.

In my fourth method, illustrated in Fig. 270, I take a piece of Norway iron, of sufficient size, draw out the shank or stem, then punch a hole, as in Fig. 270, and open out the two ends, thus getting the stem out of the solid.

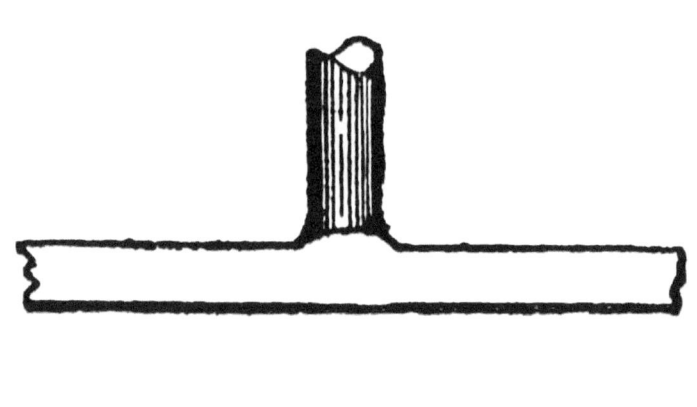

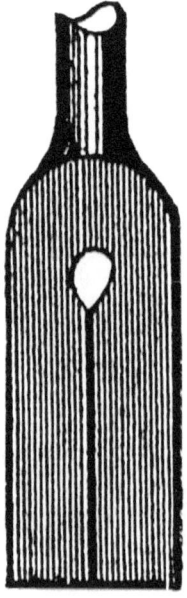

FIG. 270—THE FOURTH METHOD.

In the fifth method, as shown in Fig. 271, I take iron of suitable size, say 1 ½ x ¾ inches, make the L-shaped piece shown in the engraving, put all together, and weld them up, and complete the forging in any of the ways already mentioned.

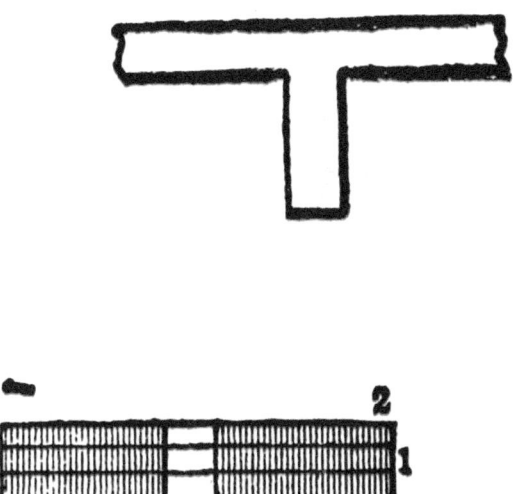

FIG. 271—THE FIFTH METHOD.

My sixth method, as illustrated in Fig. 272, is to get out a piece *B* (Fig. 272) and use the dovetail weld shown to weld in stem *A*, the stem being drawn in sideways; the corners at *x* are where the corners y will come. I split the ends of *B*, and weld in a piece *C*.

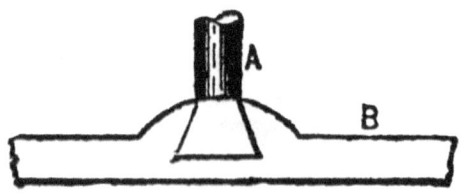

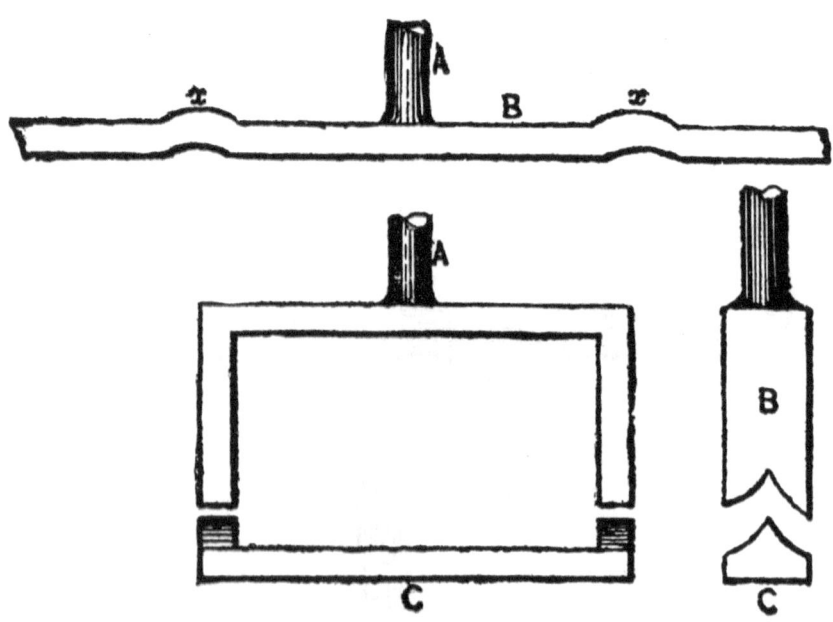

FIG. 272—THE SIXTH METHOD.

Now all these plans are good and practicable. They show six good methods of putting a shank on a flat bar of iron, and all in keeping with the general custom among skilled workmen.—*By* J. T. B.

DEFECT IN ENGINE VALVE.

The defect in many valves is owing to the fact that the cylinder exhaust port is not made wide enough for the stroke of the valve.

FIG. 273.

If the valve is worked by a rock-shaft, and there is any room to lower the pin in the rocker arm that carries the eccentric hook, the stroke of the valve can be shortened and a better proportion between the steam and exhaust ports obtained thereby. Shortening the stroke of the valve will prevent the edge of the exhaust cavity in the valve from working up close to the edge of the cylinder exhaust port, Fig. 273, and will prevent the exhaust steam from being cramped. Fig. 273 also shows part of the exhaust lap chipped from the valve, which I shall speak of presently. If the valve stroke is shortened it must have the stroke so that it will open about three-quarters of the steam port. It may be well to state that by lowering the eccentric hook pin it will be necessary to lengthen the eccentric rod, or it will throw the valve too much to the front of the cylinder, and make the port openings unequal.

As the valve has an excess of exhaust lap it would be well to take a straightedge and scriber and mark off one-third of the exhaust lap, as shown by the dotted lines in Fig. 274.

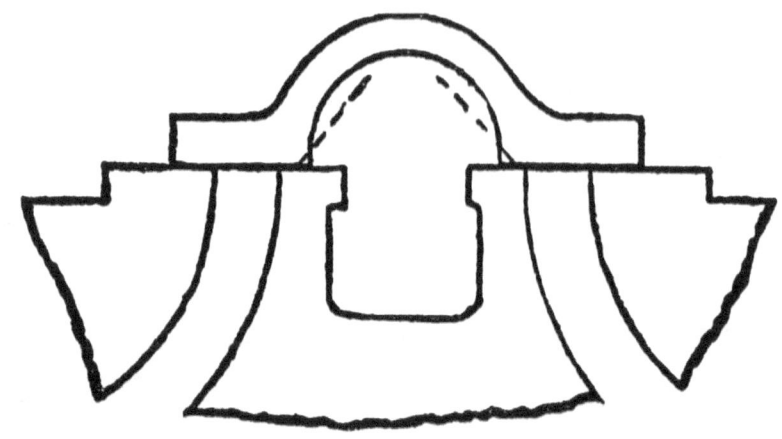

FIG. 274.

Then take a sharp cold chisel and chip off the marked edge, inclining inward to the valve at an angle of 45°, as shown in Fig. 274. By this method there will be a free and easy exhaust, and great compression prevented, even though the engine runs at a very high speed. If the valve stroke cannot be shortened, then chip out nearly one-half of the exhaust lap.—*By* Slide Valve.

CHAPTER VI.

MAKING CHAIN SWIVELS.

MAKING A LOG-CHAIN SWIVEL.

Different Methods.—Plan 1.

My way of making a log-chain swivel is as follows: I take a piece of a bar, say three-quarters of an inch square, and flatten it a little and then draw out to three-eighths of an inch round, as shown by Fig. 275.

MAKING A LOG-CHAIN SWIVEL. FIG. 275—SHOWING HOW THE PIECE IS FLATTENED AND DRAWN OUT.

I then cut it off long enough to draw out the other end the same way, leaving the center the full size, as shown by Fig. 276.

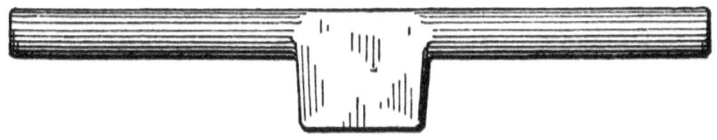

FIG. 276—SHOWING HOW THE CENTER IS FORMED.

I then punch a hole through it, and I have a mandrel shaped as by Fig. 277. I next take a welding heat, and putting the mandrel through the hole, turn up the two ends as shown in Fig. 278. After getting the ends turned up I finish it up on the mandrel with a light hammer, leaving it as shown by Fig. 279.

FIG. 277—SHOWING THE MANDREL.

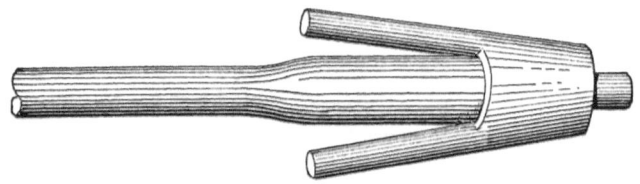

FIG. 278—SHOWING HOW THE ENDS ARE TURNED.

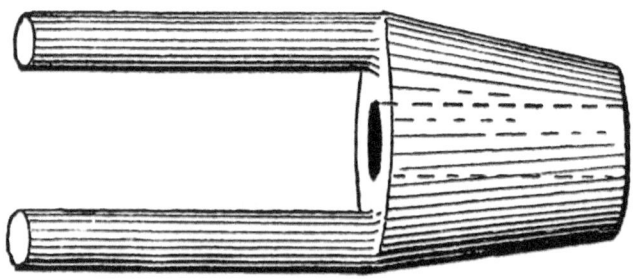

FIG. 279—SHOWING HOW THE PIECE IS FINISHED.

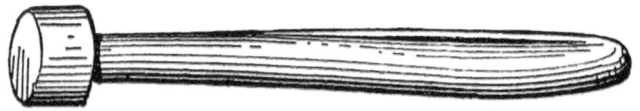

FIG. 280—SHOWING HOW THE COLLAR IS WELDED ON.

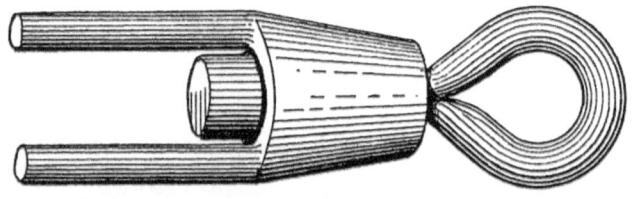

FIG. 281—SHOWING HOW THE TWO PIECES ARE JOINED.

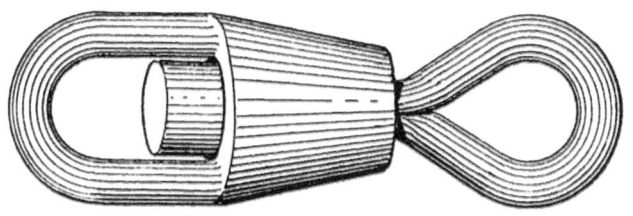

FIG. 282—SHOWING THE SWIVEL FINISHED.

I then take a piece of ⅜-inch round iron and swage it to half-round at both ends just long, enough to go through the swivel, and with half an inch to weld a collar on. I then turn it around and weld a collar round it just the size of the mandrel, as shown by Fig. 280, then heat it, drive it through the hole and open with a punch as shown by Fig. 281. I next weld the two ends together as you would a link (Fig. 282), and you will have a good swivel, as I never make them any other way, and have found them always satisfactory.—*By* John Atkins.

MAKING A LOG-CHAIN SWIVEL.

Plan 2.

To make what I call a ⅜-inch swivel, that is one for ⅜-inch log chain, I use a piece of good Swedish iron 1 ¼ inch by ¾ inch, and 4 inches long.

MAKING A LOG-CHAIN SWIVEL BY THE METHOD OF "J. H. H."
FIG. 283—SHOWING HOW THE FULLER MARKS ARE MADE.

Then I make fuller marks, as shown in Fig. 283, placing the marks just far enough apart to allow the middle part to be square.

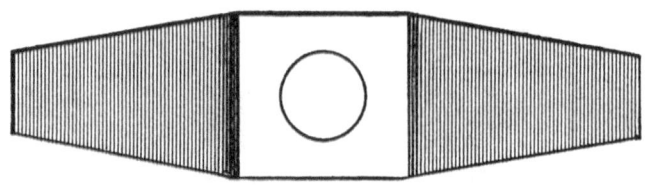

FIG. 284—SHOWING HOW THE ENDS ARE DRAWN.

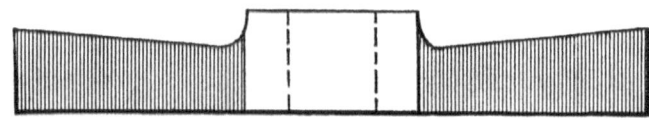

FIG. 285—ANOTHER ILLUSTRATION OF THE METHOD OF DRAWING THE ENDS.

I next punch a 5/8-inch hole in the center, and draw the ends as shown in Figs. 284 and 285. I then bend up the ends, as in Fig. 286. 290. This is made of one-inch steel (machine or Bessemer). It is drawn to five-eighths inch, and the shoulder must be nice and square if your swivel is to work easily and true.

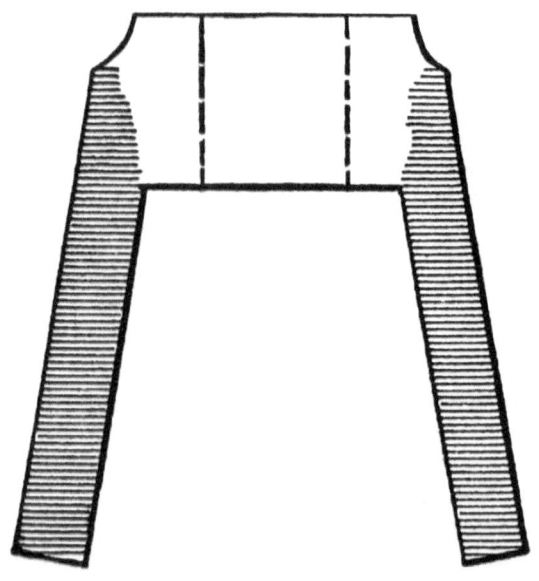

FIG. 286—SHOWING HOW THE ENDS ARE BENT UP.

After the piece has been brought to the shape shown in Fig. 287, I make the tug by taking a piece of ½ x ¾ inch square Norway iron and forging it half round for about five and one-half inches, leaving a square piece (about five-eighths of an inch) at both ends to make the head. I then bend it around, take a good soft heat, weld it up and punch the head, which must be 1 inch x ⅜ inch and round. I then forge the tug down so as to make it go in the top, as shown in Fig. 287.

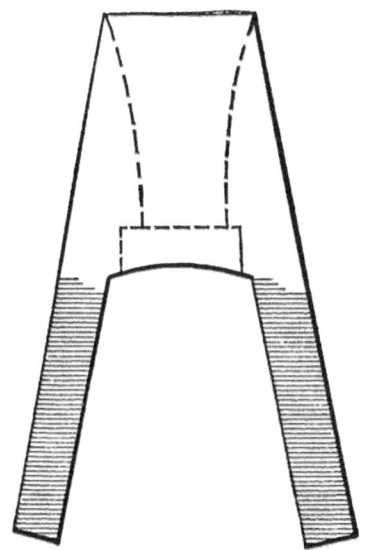
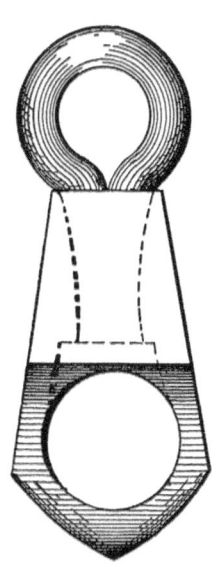

FIG. 287—SHOWING HOW THE ENDS ARE FLARED OUT.

FIG. 288—FRONT VIEW OF THE SWIVEL, SHOWING HOW THE TUG IS OPENED UP

After it has been put in, I take a very thin punch and open the tug up, as shown in Fig. 288, and weld the top together. This makes a very good swivel, and it can't freeze up in cold weather.

The dotted lines in Fig. 287 show how the end must be flared out to give the tug a chance to open.

This is what they call a bar swivel. The mandrel must be just the right size for the head, and be driven hard enough to let the head of the tug draw out of sight. After the processes described have been carried through, warm the swivel up all over, then take it to the vise to finish it. Fig. 288 is a front view of the swivel.

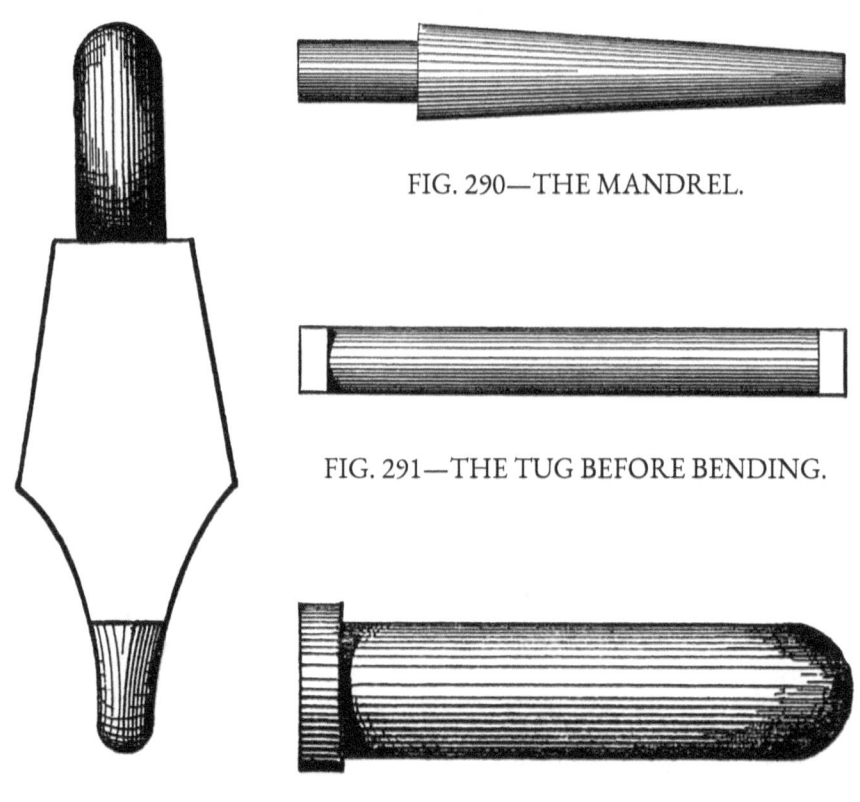

FIG. 290—THE MANDREL.

FIG. 291—THE TUG BEFORE BENDING.

FIG. 289—SIDE VIEW.

FIG. 292—THE TUG READY FOR THE SWIVEL

Fig. 289 is a side view. Fig. 290 shows the mandrel. Fig. 291 represents the tug before bending, and in Fig. 292 the tug is shown ready for the swivel.—*By* J. H. H.

MAKING A LOG-CHAIN SWIVEL

Plan 3.

To make a log-chain swivel cheaply and quickly, take a piece of round iron one size larger than the wire in the chain, flatten about six inches quite thin and turn at a right angle and a square corner as in Fig. 293.

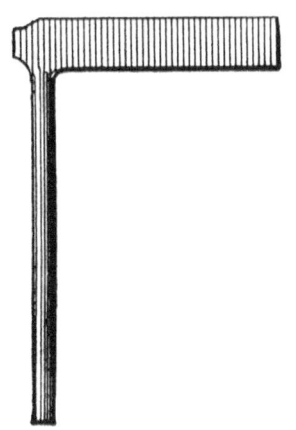 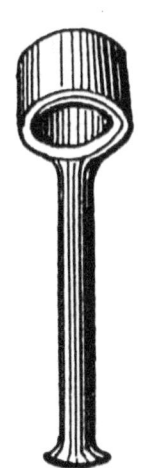

MAKING A LOG-CHAIN SWIVEL. FIG. 293—SHOWING THE IRON FLATTENED AND TURNED

FIG. 294—SHOWING HOW THE HOLE IS MADE FOR THE SWIVEL

Then chamfer the out corner with a flat punch to get a lap, then commence at the end of the flat part and roll it up to the round, leaving a hole of about half an inch for the swivel for a three-eighths inch chain, as shown in Fig. 294.

Then cut off the round about five inches from the corner, scarf the end short, thin and wide, and turn it on to the other side. Next take hold of the bow thus made, take a good weld and weld it on the horn as in Fig. 295.

FIG. 295—SHOWING HOW THE BOW IS MADE.

FIG. 296—SHOWING HOW THE OTHER PIECE IS BENT AND CUT OFF.

Hold your hand higher than the horn and you will get a tapering hole. Then take a piece of the same round iron, bend over about three inches, flat them together a little, and turn and weld a ring for a head, finish it up over the corner of the anvil and cut it off so that both ends are of the same length, as in Fig. 296.

FIG. 297—SHOWING HOW THE TWO PIECES ARE JOINED.

Heat the swivel, drive it through, open the ends as in Fig. 297, and turn and weld as you would a link. Then heat the whole and work loose, and the job is completed. Use a light hammer when welding.—*By* Home.

MAKING A SWIVEL FOR A LOG CHAIN.

Plan 4.

To make a swivel for log chains, I first forge the piece shown in Fig. 298, making the hole by punching.

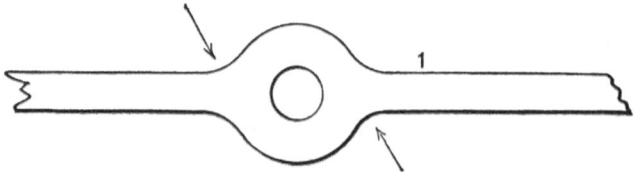

MAKING A SWIVEL FOR A LOG CHAIN BY THE METHOD OF "C. E. B."
FIG. 298—SHOWING THE PIECE AS FORGED AND BEFORE BENDING.

The stem, shown in Fig. 299, should be made large enough to fit the hole in Fig. 298 neatly. A is a washer used when fitting the stem in the hole.

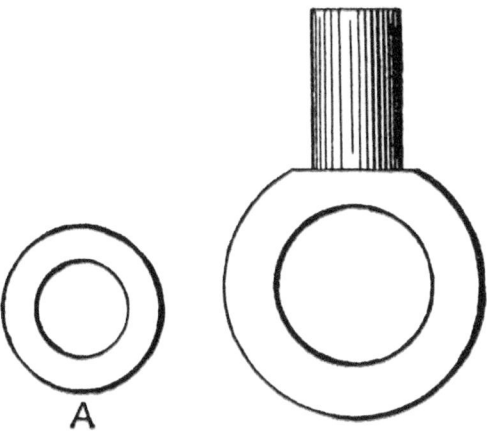

FIG. 299—SHOWING THE STEM AND WASHER.

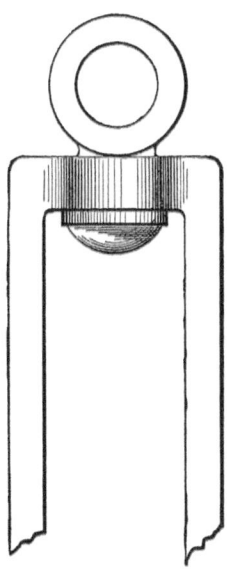

FIG. 300—SHOWING THE RESULT AFTER RIVETING.

The piece shown in Fig. 298 is then bent as indicated by the arrows, and the iron is kept cool excepting where the bends are made. The ends are left open. The next step is to heat the end of the stem, Fig. 299, to a cherry red, and rivet it to the piece shown in Fig. 298.

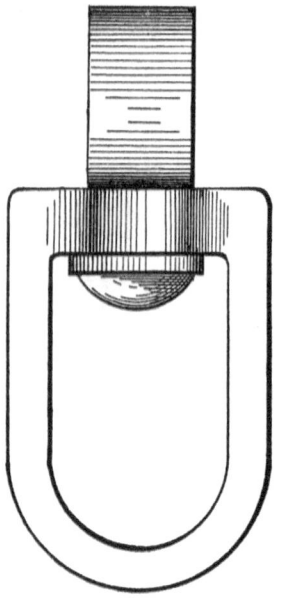

FIG. 301—SHOWING THE JOB COMPLETED.

In doing this use a vise and punch. The punch I use for the purpose is countersunk for a small space on the bottom. The result of this riveting is represented in Fig. 300, and by bending the points together and welding the job is completed as indicated in Fig. 301.—*By* C. E. B.

MAKING A SWIVEL FOR A LOG CHAIN.

PLAN 5.

In making a swivel to a log chain, I take a piece of good iron and forge it as at *A B*, Fig. 302, which are edge and top irons, *B* being a hole.

I next forge a piece, *C*, shown in Fig. 303, and put it through the hole in *A*, and open out the split to form the eye, and then weld up the ends of *A*, and the job is complete, as illustrated in Fig. 304.—*By* H. K.

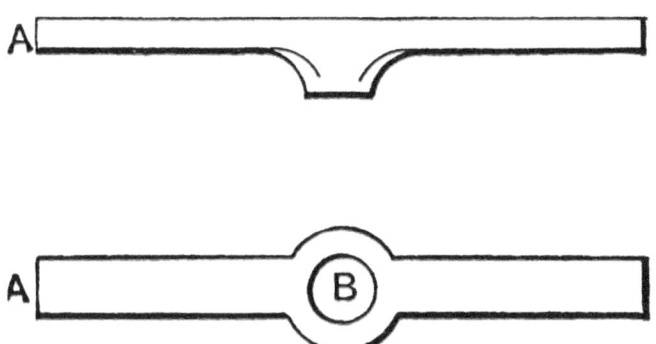

MAKING A SWIVEL FOR A LOG CHAIN, AS DONE BY "H. K."
FIG. 302—SHOWING THE EDGE AND TOP IRONS.

FIG. 303—SHOWING THE PIECE USED FOR THE EYE.

FIG. 304—SHOWING THE FINISHED JOB.

MAKING A SWIVEL FOR A LOG CHAIN.

Plan 6.

My way of making a swivel for a log chain is to: 1st. Make an eye punch of one and one-eighth inch round iron, as shown in Fig. 305.

MAKING A SWIVEL FOR A LOG CHAIN.
FIG. 305—THE EYE PUNCH.

2d. Take a piece of good iron 1 ½ x ½ inch, four inches long. 3d. Fuller and draw ends as shown in Fig. 306. 4th. Bend the ends, as illustrated in Fig. 307.

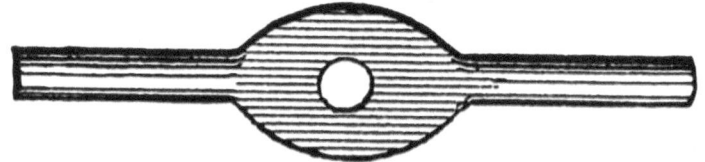

FIG. 306—SHOWING HOW THE ENDS OF THE PIECE ARE FULLERED AND DRAWN.

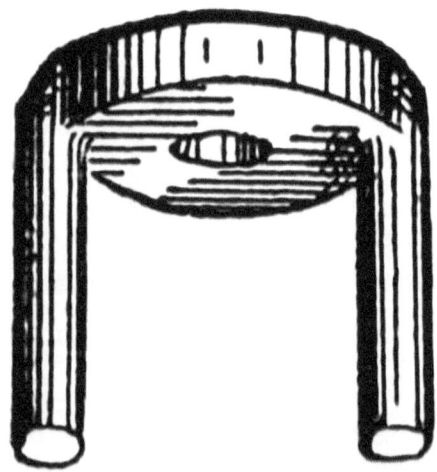

FIG. 307—SHOWING HOW THE ENDS ARE BENT.

5th. Put in eye punch and work down, as in Fig. 308.

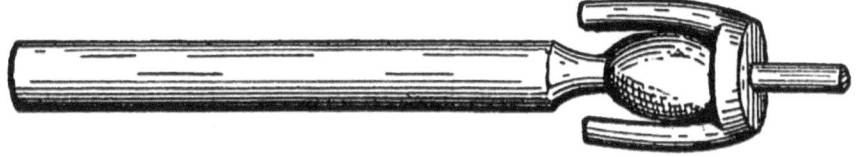

FIG. 308—SHOWING HOW THE EYE PUNCH IS USED.

6th. Make the eye as shown in Fig. 309. 7th. Square the top end of the eye and make the washer three-eighths of an inch thick, with a square hole.

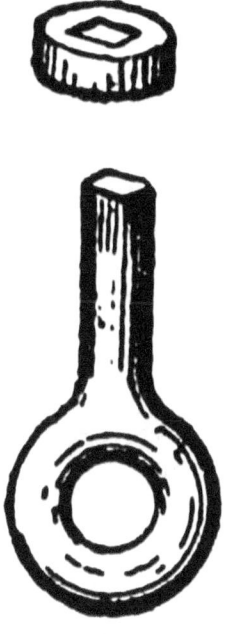

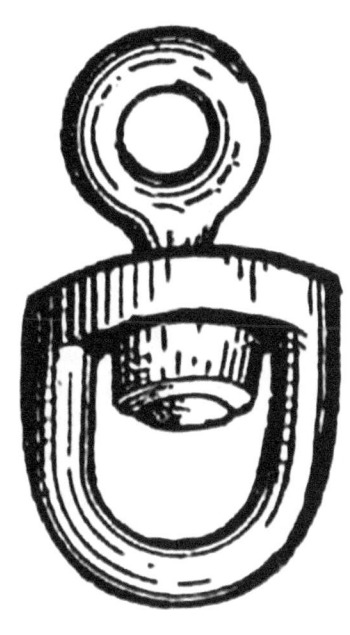

FIG. 309—SHOWING HOW THE EYE IS MADE

FIG. 310—THE FINISHED SWIVEL.

Heat the end of the eye and washer, put the eye in the vise, put on the swivel, then put on the washer and rivet. Then weld the swivel. When finished (see Fig. 310) it will not be more than four inches long.—*By A. S., Jr.*

MAKING A SWIVEL FOR A LOG CHAIN.

Plan 7.

I think the best way is to make the link of square iron, drawing off the ends, as shown at 1 in Fig. 311, and leaving a heavy center in which the eye is to be punched. When the bending is to be done use the tool marked 2, Fig. 311.

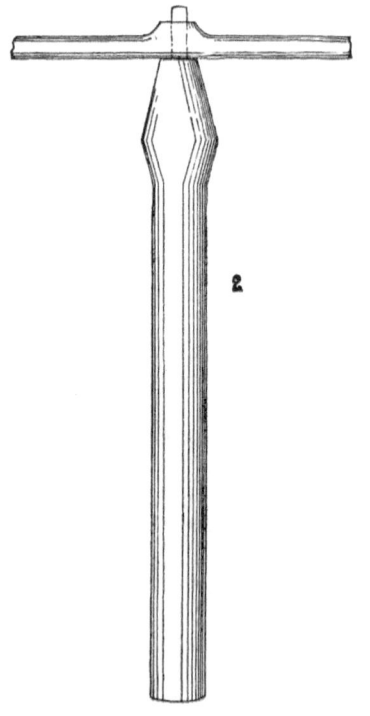

MAKING A SWIVEL BY THE METHOD OF "J. E. N."
FIG. 311—SHOWING THE SHAPE OF THE ENDS AND THE TOOL USED IN BENDING

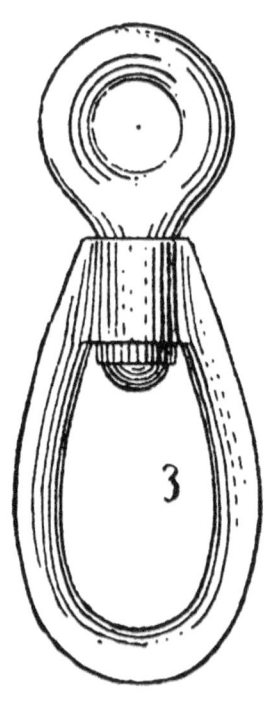

FIG. 312—THE SWIVEL COMPLETED.

Then finish on a swage, rivet the swivel eye in hob, weld the ends of the swivel link, and while hot turn the eye around a few times. Then cool off, and the result will be a swivel as shown in Fig. 312.—By J. E. N.

CHAPTER VII.

PLOW WORK.

POINTS IN PLOW WORK.
No. 1.

My method of putting plowshares on stubble plows is as follows: I commence by preparing the landside, as in Fig. 313, and attaching it on the plow in the proper place.

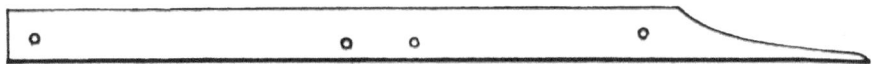

POINTS IN PLOW WORK. FIG. 313—SHOWING THE LANDSIDE.

I next cut out the steel, as in Fig. 314, then heat the part from A to B, and drive it over so that when the back of the steel is against the moldboard the line $A B$ will be on a direct line with the landside. I next plate it out to a thin edge from heel to point, and then heat it evenly all over so that it will bend easily to the blows of the hammer.

FIG. 314—THE STEEL.

In preparing this part of the lay to fit the landside and moldboard, the smith should be very careful to leave no hammer marks on the front side, and he should also be careful to get a good fit, as that is much easier at this stage than when the part is welded on the landside. When this piece of steel is fitted as I have described, attach it to the plow by fastening it at the heel to the brace, and in front by fastening it to the landside with a clamp made after the pattern shown in Fig. 315. This clamp should be made so that after being put on at the point it will be driven tight when about two-thirds of the way upon the lay, thus holding the steel fast to the landside so that it will weld easily.

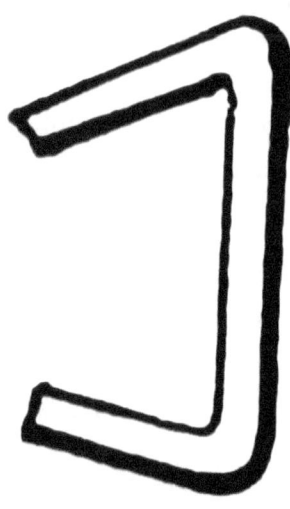

FIG. 315—THE CLAMP.

The next thing to be done is to unfasten the lay from the plow. Prepare a good fire and weld quickly, being careful not to let the piece slip or disturb it any more than is necessary. When it is welded to the top, and before finishing the point, lay a piece of steel from the edge of the point back about two or three inches; this will make the part last much longer. Next finish up the edge and point and edge of the lay, and if it has been held in place you will have a perfect fit.

When it is desired to put on a point, in preparing the point I forge one end of the steel flat and draw it to a thin edge, the other I draw to a point, as in Fig. 316, and then double it so the pointed edge will go on the bottom of the landside. I next weld both sides down solid, draw out to required thickness, cut off the edge of the point at the same angle as the lay, finish up and the point is complete.

FIG. 316—THE POINT.

Sometimes a lay is brought to the shop that is worn very thin, and some are worn through at the top, near the moldboard. In this case I forge the point the same as I would in the one just described, but leaving the flat side long enough to reach the top of the lay. I then bore a hole near the top through both the piece of steel and the lay and fasten them with a rivet. I next heat the piece of steel where it is to be doubled over, letting the pointed piece go on the bottom of the landside, weld up as high as is convenient, and finish as before.

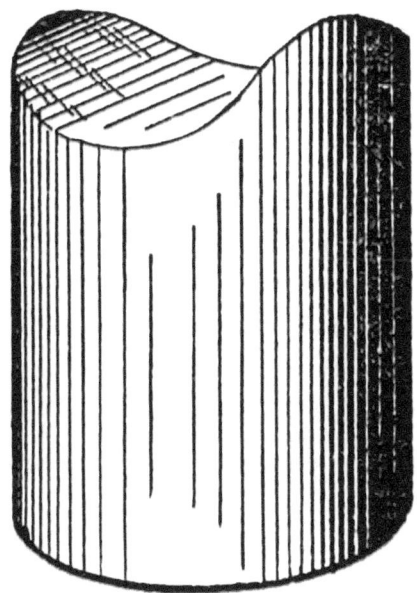

FIG. 317—THE BLOCK USED IN COOLING THE LAY.

For the purpose of cooling a lay easily and with more rapidity, I use a wooden block about eighteen inches in diameter and two feet long. I square one end, place it upon the floor, and saw the other end to the shape of a half circle, as shown in Fig. 317. This cavity is lined with iron so the wood will not burn. When the lay is heated I put it upon the opening and place a piece of round iron about three feet long and two inches in diameter upon the part that is to be shaped, one end of the piece being held in my hand; the helper then strikes upon the round piece with a heavy hammer and the lay will bend evenly without leaving marks.—*By* T. M. S.

POINTS IN PLOW WORK.
No. 2.

I always weld my points on the top instead of the bottom. My reason for this is, that the share always wears thin above the point, and by drawing your point tapering, leaving the back thick, and drawing the edge that extends on the face of the share thin and wide, you make the weak places strong. All smiths know that the throat, especially where it is welded next to the bar, wears thin, and unless you place your point on the top side, it is soon worn through, and the share must be thrown aside. In drawing the shin piece, the piece that is welded from the point of the share to the point of the mold should be drawn to a feather edge where it extends on to the wing and left heavy on the back, so that when it is welded on it will scour off and become smooth. I have seen many shares that would not scour simply from the fact that the edge had been left thick, and in welding the smith made the wing sink in, and left a place for the dirt to stick.—By W. F. S.

SOME MORE POINTS ABOUT PLOWS.

That bolts are among the most important things when it comes to repairing plows, I shall endeavor to show in the following remarks.

In all the plows made in factories, that is the steel plows, with very few exceptions, the round counter-sink key head is used, or else a square neck round countersink head, which is very little better than the key head. It seems to me that plow factories lose considerable profit on this item of bolts.

My plan consists in simply punching a square hole, tapering to fit a square plug head bolt. I hear some one ask how would I punch a tapering hole. I will

FIG. 316—THE POINT.

Sometimes a lay is brought to the shop that is worn very thin, and some are worn through at the top, near the moldboard. In this case I forge the point the same as I would in the one just described, but leaving the flat side long enough to reach the top of the lay. I then bore a hole near the top through both the piece of steel and the lay and fasten them with a rivet. I next heat the piece of steel where it is to be doubled over, letting the pointed piece go on the bottom of the landside, weld up as high as is convenient, and finish as before.

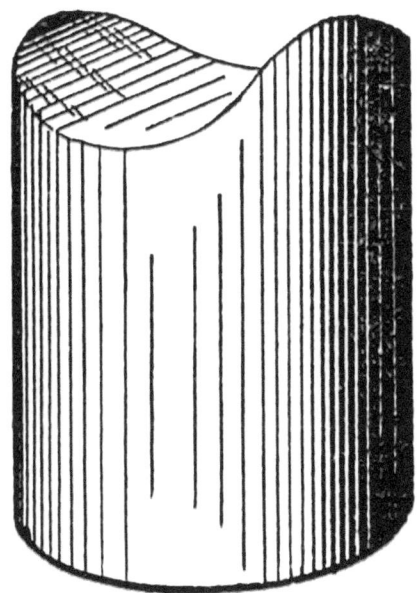

FIG. 317—THE BLOCK USED IN COOLING THE LAY.

For the purpose of cooling a lay easily and with more rapidity, I use a wooden block about eighteen inches in diameter and two feet long. I square one end, place it upon the floor, and saw the other end to the shape of a half circle, as shown in Fig. 317. This cavity is lined with iron so the wood will not burn. When the lay is heated I put it upon the opening and place a piece of round iron about three feet long and two inches in diameter upon the part that is to be shaped, one end of the piece being held in my hand; the helper then strikes upon the round piece with a heavy hammer and the lay will bend evenly without leaving marks.—*By* T. M. S.

POINTS IN PLOW WORK.
No. 2.

I always weld my points on the top instead of the bottom. My reason for this is, that the share always wears thin above the point, and by drawing your point tapering, leaving the back thick, and drawing the edge that extends on the face of the share thin and wide, you make the weak places strong. All smiths know that the throat, especially where it is welded next to the bar, wears thin, and unless you place your point on the top side, it is soon worn through, and the share must be thrown aside. In drawing the shin piece, the piece that is welded from the point of the share to the point of the mold should be drawn to a feather edge where it extends on to the wing and left heavy on the back, so that when it is welded on it will scour off and become smooth. I have seen many shares that would not scour simply from the fact that the edge had been left thick, and in welding the smith made the wing sink in, and left a place for the dirt to stick.—By W. F. S.

SOME MORE POINTS ABOUT PLOWS.

That bolts are among the most important things when it comes to repairing plows, I shall endeavor to show in the following remarks.

In all the plows made in factories, that is the steel plows, with very few exceptions, the round counter-sink key head is used, or else a square neck round countersink head, which is very little better than the key head. It seems to me that plow factories lose considerable profit on this item of bolts.

My plan consists in simply punching a square hole, tapering to fit a square plug head bolt. I hear some one ask how would I punch a tapering hole. I will tell you how I do it. The size of bolt I use for a bar is seven-sixteenths and half inch. Make a square punch a trifle larger than the bolt, so it will go in without driving and will not spoil the threads, then make the die, say for a seven-sixteenths inch bolt, three-eighths of an inch, set it so the punch will pass through the center, now try it and see if it does not make a taper hole fitting a plug-head plow bolt. Make any other size punch and die in about the same proportions to get the taper. You can punch every hole in the mould the same way. Sometimes the corners of the hole will not be cleanly cut out, and in that case take a square pointed punch tapered and tempered, set it in the hole and hit it once or twice.

I think I have shown where two expensive operations with the vise and drill can be saved, and in doing this it is necessary to have only one variety of bolts to use. Then when it comes to repairing the same there is no turning of bolts in the hole. If it is so rusted that the nut won't turn it is better to twist it off and put in a new bolt, at a cost of five or ten cents, than to spend a half hour trying to get the old one out, and then have to charge for the time and bolt beside.—*By* Dot.

MAKING A PLOW LAY.

I will describe my method of making a plow lay:

If a new stub on landside is to be made forge it out first. Sometimes the old one can be used by welding a piece on the point and letting it run upon the top of the old stub. After the stub is bolted on, lay the slab of steel on the plow and mark the bevel, then lay the block of a two-inch square on the steel, and mark as shown in Fig. 318 of the accompanying illustrations.

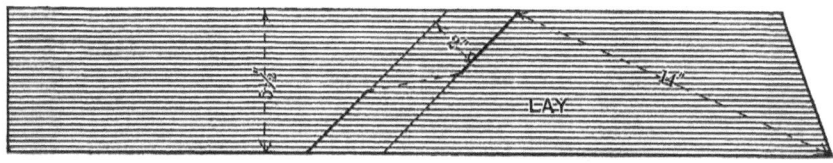

MAKING A PLOW LAY.
FIG. 318—SHOWING HOW TO MARK THE STEEL.

Next heat and turn the point and draw and cool the lay so that when the plow rests on the floor the edge will touch over the entire length of the lay. Next cut out a small piece of steel to lay on the bottom of the point, and punch a hole through the point of the lay, the landside and the small piece of steel, and rivet them all together. Then take off together the entire landside and lay, weld the point first, then take off the long landside and weld the upper end of the stub to the lay, and weld the middle last.

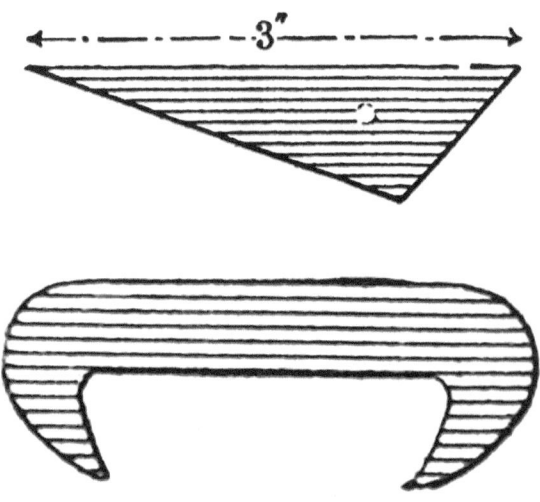

FIG. 319—SHOWING CLAMP FOR HOLDING WHILE HEATING THE POINT. FIG. 320—SHOWING POINT USED FOR BOTTOM OF LAY.

Never try to weld from the point up, for the steel will curl or spring up from the stub and prevent your getting a good weld. While heating the point hold it by means of the clamp shown in Fig. 319. Fig. 320 represents the point used for the bottom of the lay. I use borax for a flux.—*By* B. N. S.

LAYING A PLOW.

To describe how I lay a plow, I will begin by calling attention to Fig. 321, which shows how a plow generally looks when brought to the smith to be laid and pointed. There is usually a hole worn in the plow back of the point in the throat, as in 1 in Fig. 321.

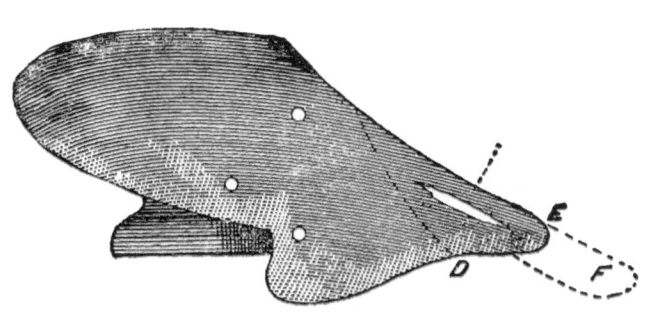

LAYING A PLOW BY THE METHOD OF "J. O. H."
FIG. 321—SHOWING THE PLOW BEFORE LAYING AND POINTING.

I forge out a piece of steel of the shape shown in Fig. 322, making it very thick on the side from *A* to *B* and then from *A* to *C*; I then clamp it to the plow with tongs, as in *D*, Fig. 321, and as shown by the dotted lines. I next heat at *E*, weld rapidly, take off the tongs, bend the part F under, then put on plenty of borax and weld clear up. The plow is now solid and ready for the lay.

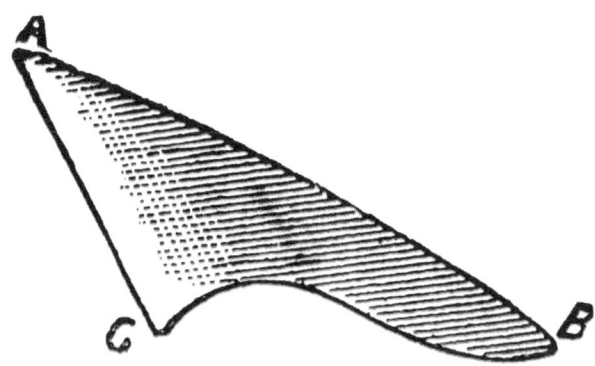

FIG. 322—SHOWING THE PIECE TO BE CLAMPED TO THE PLOW.

I forge out a piece of lay steel of the shape shown in Fig. 323. It is simply doubled up at the point *G*; I forge then from *H* to *I* and *J*, and upset at *H* very heavy, to bring out the worn-off corner of the plow; I then lay it on the bottom of the plow as shown by the dotted lines in Fig. 324, clamp it on with tongs at *K*, put a heat on the point, weld rapidly, then take off the tongs and weld up solid as far back as the lay goes.

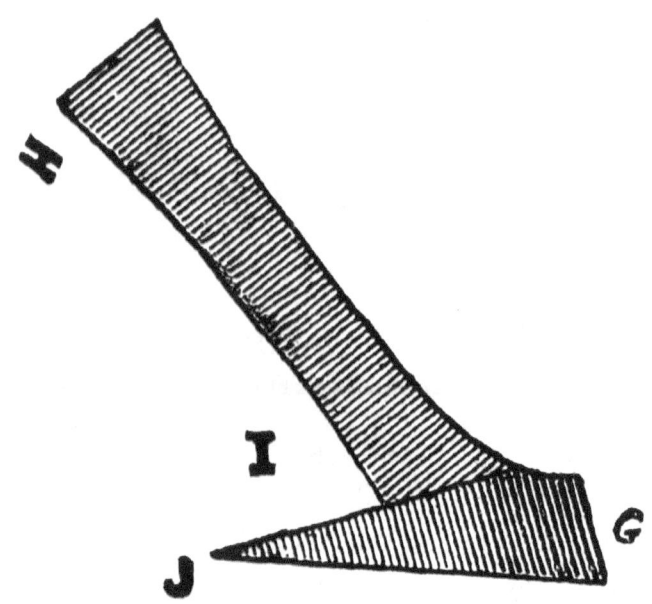

FIG. 323—SHOWING THE SHAPE OF THE LAY STEEL.

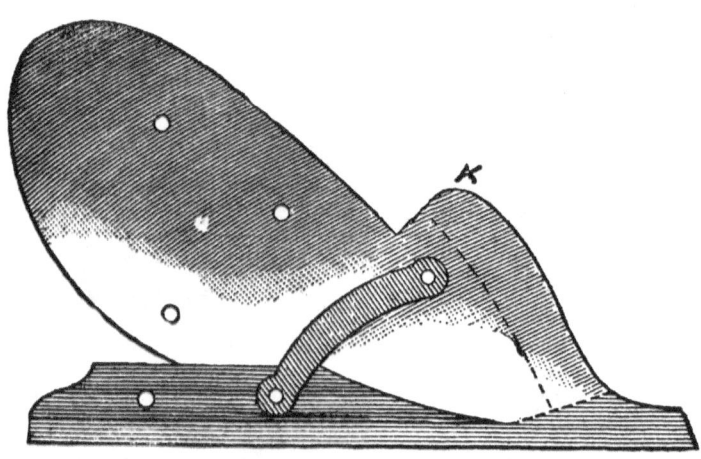

FIG. 324—SHOWING HOW THE STEEL IS LAID ON THE PLOW.

The heat should be put at the back part of the lay, along the dotted line in Fig. 324, to make a good weld back there. If the job is properly done the share will be lengthened one and a half to one and three-quarters inches.

Fig. 325 shows how the plow looks after laying, and the dotted line indicates its appearance before the laying. Most of the laying I do is on Diamond plows. In laying a new share, great care is required to keep it in shape.—*By* J. O. H.

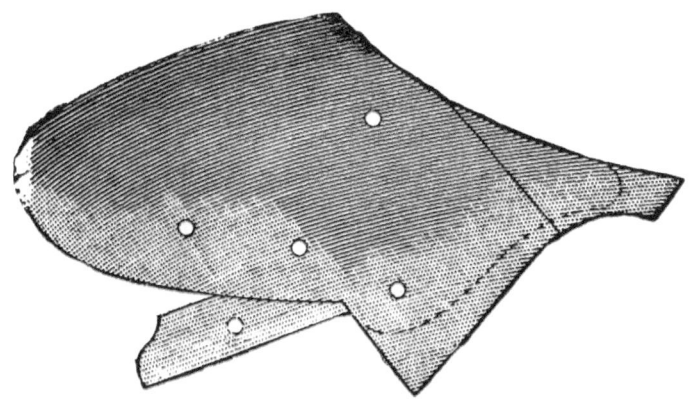

FIG. 325—SHOWING THE JOB COMPLETED.

POLISHING PLOW LAYS AND CULTIVATOR SHOVELS.

Perhaps some of your readers would like to know a good way to polish plow lays and cultivator shovels. After pointing shovels and new or old lays I always grind them bright on a solid emery wheel about twelve inches in diameter, with a two-inch face. I then put them on the felt wheel and finish them off so that they look as if they had just come from the factory. The chief point to keep in mind in using emery wheels for a smooth surface like a plow lay is to have a wheel that is soft and elastic. You can never get a fine finish from a hard, solid stone.—*By* "Shovels."

LAYING A PLOW.
Plan 1.

In laying a common plowshare I make my lay of German plow steel about five-sixteenths of an inch thick on one edge, and as thin as I can hammer it on the other. I then sharpen the plowshare as sharp as the thin edge of the lay; then I strike all the point in separate heats. If I have no striker, I weld about three inches at a heat, being careful to get the back edge thoroughly welded down and level with the surface of the balance of the share. If the outer edge of the share is not thoroughly welded, this can be easily done by taking a heat on it while drawing the edge out sharp. If I have a striker, I weld about four inches at a heat, having the striker strike directly over where the edge of the share meets the lay, thus welding the edge of the share to the lay, while I weld the back edge of the lay to the share with the hand hammer at the same time.

The greatest trouble with most smiths in laying plows is that they leave their lay too thick at the back edge where it laps over the share, so the share gets too hot and burns before the lay gets hot enough to weld. I use common finely pulverized borax for a welding compound. A good way to test your heat is, when the lay is about to come to a welding heat, just lightly remove the burning coal from the top of the lay and lightly tap its edge along as it welds, keeping up a blast at the same time. This way of doing not only shows when the lay is in a condition to weld, but the welding is actually commenced in the fire, the share and lay thus beginning to unite as soon as it begins to come to a weld, and therefore is less liable to burn than if it were not settled by the light blows of the poker. I always leave the out edge of my lay as thick as possible until all is thoroughly welded. Then the plow is ready for the point; for points I use $1 \frac{1}{2} \times \frac{1}{2}$ inch German steel.

I get my point out with the back end as wide as the base it is for, and as thick as the base will permit. I let my point extend back from two to five inches, or sufficient to give the base the necessary strength at the point. I strike the point at separate heats, and when the point is thoroughly welded I work it out well in the throat with the pene of my hammer or fuller, then begin at the point to draw the shoe out, drawing it with a gradual slope from the back edge of the lay to the required edge, keeping the edge on a perfect level with the bar, and being careful to leave the heel of the share and the corners of the point on a straight line, sighting from the heel of the share to the point.—*By* J. McM.

LAYING A PLOW.
Plan 2.

To new lay steel plowshares, I select a piece of spring steel and heat one end to a high borax heat, then strike it with my hammer, and if it flies to pieces I put it aside. I don't use high steel. In preparing my steel I partially weld a strip of hoop iron on the edge, leaving it so that where it is scarfed or chamfered the iron will be full with the edge of the steel. I then chamfer the share and weld as usual. It is surer to weld when iron is between the parts.—*By* J. U. C.

LAYING A PLOW.
Plan 3.

Repairing plows forms a major part of the blacksmith's business in the Pacific States from the first of October until the following May. I will therefore give my plan of "laying a plow," and I hope by this means to draw the craft into a discussion of the subject, to our mutual benefit.

I will take a sixteen-inch plow for example. First cut a bar of iron 2 ½ x ½ inch and about thirty inches long; cut on a bevel at one end to save hammering; take a heat on the beveled end; draw as in Fig. 326, from shin to point about ten inches long, and punch a one-quarter inch hole, as at *A*.

Second: Clamp the landside thus prepared upon the plow standard in the proper position (the plow being inverted), and with a slate pencil mark for the holes, which, when drilled, countersunk and squared to fit the bolts, can be bolted into place upon the plow.

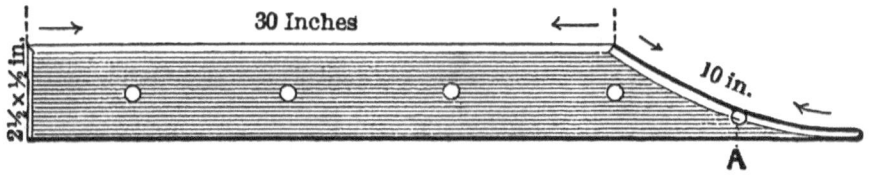

LAYING A PLOW BY "HAND HAMMER'S" METHOD. FIG. 326—
SHOWING THE IRON BAR DRAWN FROM SHIN TO POINT.

Third: Take a slab of steel 6 x ¼ inches, lay it on the plow, parallel with and against the lower edge of the moldboard, and with a pencil draw a line across the steel against the landside. This gives the length and angle at which to cut the steel, as at *b*, Fig. 327.

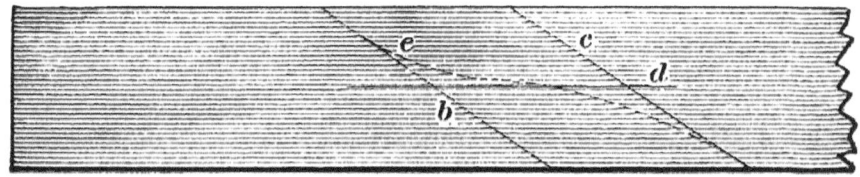

FIG. 327—SHOWING THE METHOD OF CUTTING THE STEEL.

Then lay the blade of a square two inches wide upon the steel, parallel with the line *b*, and draw a second line on the opposite edge of the square, which gives us the line *c*, parallel with and two inches distant from line *b*.

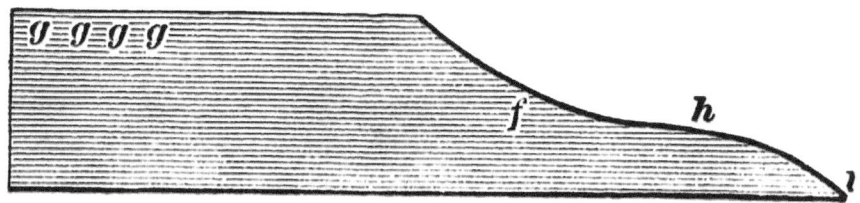

FIG. 328—SHOWING THE PATTERN.

Now find the center of the cross section and draw the line *d*, then trace with a pencil as indicated by the dotted line *e*, and upon this line heat and cut off, which gives you the pattern Fig. 328. Next heat at *f*, Fig. 328, as hot as the steel will bear, and grasping the pattern with both hands at *g, g, g, g*, strike the point *h* upon the anvil a few sharp blows, which will give you the "shape," Fig. 329, after which draw to an edge, then heat as hot as the "law allows" the entire length and bend over a mandrel (lying down), the helper holding a sledge on the back of the pattern while you hammer along the edge until the curve is right to fit upon the landside and brace, and along the moldboard where it is designed to go. When cool place the pattern in position and mark for the holes 1, 2, 3, 4, as shown in Fig. 329.

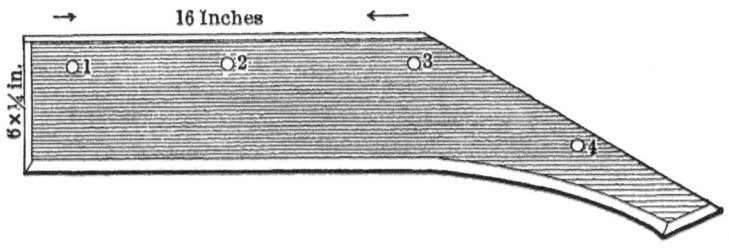

FIG. 329— H WING THE "SHAPE."

After these holes are drilled, countersunk, and fitted to the bolts, place the parts in position as in Fig. 330, and fasten with a bolt at *I*, and a rivet at *J*.

Now you are ready for the last, though by no means least important or difficult operation, *i. e.*, welding, for if you fail in this all is lost.

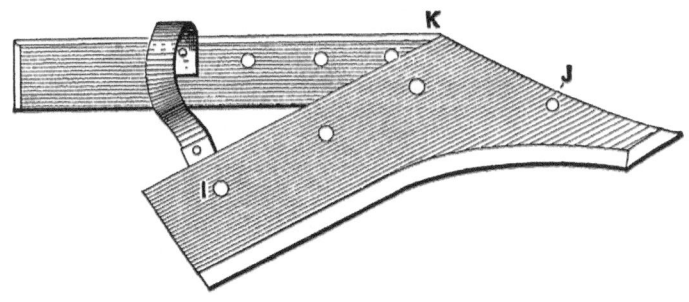

FIG. 330—SHOWING THE PARTS PLACED IN POSITION AND FASTENED.

Finally: Place the work in a roomy, clean fire with the landside flat down, and heat red at the shin; then take it to the anvil and hammer it down close; return it to the fire, apply plenty of borax, and heat to a good welding heat, giving plenty of time to "soak;" then place it upon the anvil and have your helper catch hold of it at the top of the shin at *K*, Fig. 330, with the "pick-ups" and squeeze firmly to the landside while you weld with the hammer from the tongs down to the rivet. Next take a heat at the top of the shin where the tongs prevented welding and weld it down, after which weld from rivet to point. The slender point from *h* to *l*, Fig. 328, must be turned under; then weld up, which gives strength and shape to the point. Now straighten up the edge, and temper the bolt on the plow.—*By* Hand Hammer.

LAYING, HARDENING AND TEMPERING PLOWS.

I will try to give my practical experience in laying, hardening and tempering the Casaday sulky plowshare; a plow generally used in Texas for breaking old land, and sod as well. When the shares are new they cut from twelve to fourteen inches, and are five and one-half inches wide, measuring from twenty to twenty-five inches from heel to point on the edge. When they are brought to the shop for laying, they resemble Fig. 331 of the accompanying illustrations, measuring from three and one-half to four inches in width.

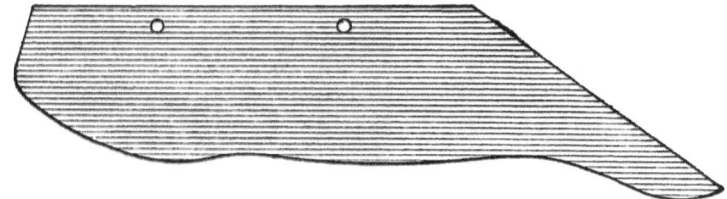

FIG. 331—SHOWING THE APPEARANCE OF THE SHARE WHEN BROUGHT TO THE SHOP FOR LAYING.

The first thing I do is to lay a piece of iron on the point, which, when shaped looks like Fig. 332.

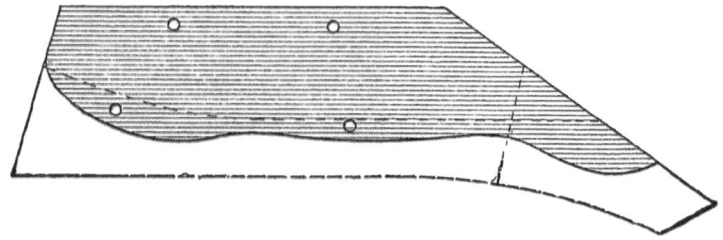

FIG. 332—SHOWING HOW THE IRON IS LAID ON.

This makes the point of sufficient length and strength to receive the lay, which I make of German or hammered steel, one and one-quarter inches wide by three-eighths of an inch thick, and shaped as in Fig. 333. The heel of the lay is upset to make it heavier and wider in order to supply the deficiency of metal at the heel of the share. I then draw the upper or inside edge of the lay to about one-eighth of an inch, as I do also the edge of the share, and next drill two small holes in the share; one at the heel, the other about midway between the heel and the point of the share.

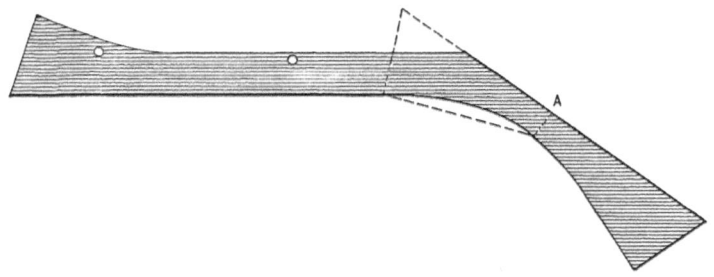

FIG. 333—SHOWING HOW THE LAY IS SHAPED.

I then place the lay in position on the under side of the share, mark and drill the lay, and rivet it on. This will hold it in position. I then put it in the fire and heat until I can bring the lay and the share close together, and then turn the lay back from A, the top of the share, as shown at A, Fig. 333, the object being to thicken the steel, which in many cases is worn quite thin, and requires this extra metal to draw and shape the point. The dotted lines show how the broad or fan-tail points of the lay fit the share when turned back. I next weld from point to heel, and if when welded the back of the share will not fit the moldboard, I heat the whole share, and while it is hot set it up edgewise on the anvil, and strike it on the edge, which will bring it straight. I then begin to shape it at the point, and draw the edge, using the pene of the hammer, or, if the face is used, I allow the edge to project over the round edge of the anvil, to prevent stretching on the edge, which would cause it to curve on the back.

When finished and properly shaped it will look like Fig. 334, and be from one-half to three-quarters of an inch wider than the original share. Before drawing it to an edge, I carefully examine it to see if there are any skips or failures, and if such are found, heat, raise the edge, and insert a thin piece of hoop iron, allowing it to project slightly beyond the edge of the skip. Take a light borax heat and it will stick.

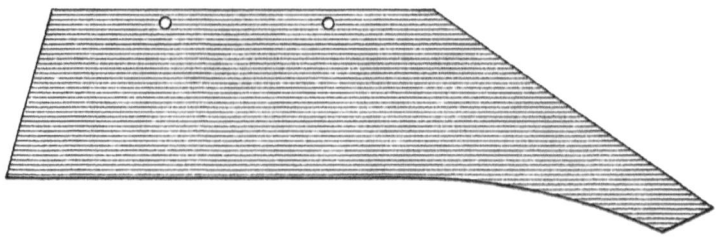

FIG. 334—SHOWING THE SHARE COMPLETED.

With the share finished the next thing in order is hardening and tempering, which if done in the ordinary forge of the country shop requires practice for success. First have a good clean fire of well coked coal, keep up a regular, steady blast as if taking a. welding heat, put the point in first, allow it to get nearly red, as it is the thickest, move it through the fire gradually to the heel, and continue to pass it back and forward through the fire until you have an even heat from point to heel about an inch back from the edge. When you have the proper heat, have at hand a trough two and one-half feet long, and holding the share firmly in tongs immerse the edge in the water for an instant. Raise the edge or heel out first, allowing the point to remain a second longer because it has more heat than the edge, but taking care to have heat enough remain in the share to draw the temper to a blue. Have a brick with which to rub the edge, that you may be able to see when you have the proper temper. If there should not be heat enough in the share to do this, hold it over the fire again. Be careful to draw the temper as evenly as possible, for if it has hard and soft places it will wear into scallops.—*By* D. W. C. H.

LAYING PLOWS.

I have been laying plows more or less for sixteen years, and in describing my method I will first deal with the slip share. I first make a lip as shown in Fig. 335. This lip is made of iron one-half an inch thick, and of proper width, forged down to the shape shown, with a bevel for right or left as desired. The joint *A* must be fitted snugly to the plow, so that when bolted on it will be level with the bar on the bottom, and will neither work up nor down at the point. In drilling the hole in the lip, care must be taken to have the bolt draw the joint together.

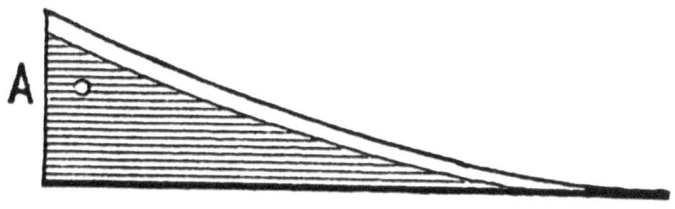

LAYING PLOWS BY THE METHOD OF "R. W. H."
FIG. 335—THE LIP.

When finished, bolt on the plow, and swing the plow up by the end of the beam, letting the handles rest on the shop floor. Cut the steel as in Fig. 336, using five-inch steel for twelve-inch plows, five and one-half inch for fourteen-inch, and six-inch for sixteen-inch plows.

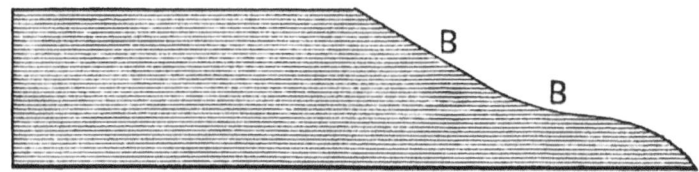

FIG. 336—SHOWING HOW THE STEEL IS CUT.

Heat the steel at *BB* hot as the law allows, and bring the point around to the proper angle, and to the shape of Fig. 337. The edges of the shares should be drawn with the face to the anvil, making the bevel underneath. Cut and fit to the lip and mold-board as snugly as possible. It is a good plan to have the point of the share longer than necessary, so it can be turned under the back when beginning to weld on. When your lay is in the exact place, mark the boltholes and drill and countersink, then bolt the share to the plow with a crossbar bolt. Take the landside, crossbar, and share all off together, just as you would a solid bar, and with a nice, large, clear fire weld up the two first heats, then unbolt the share and lip, and take off and finish welding to the top; now bolt on the landside and refit the share, if you have got it out of shape in making the last welds.

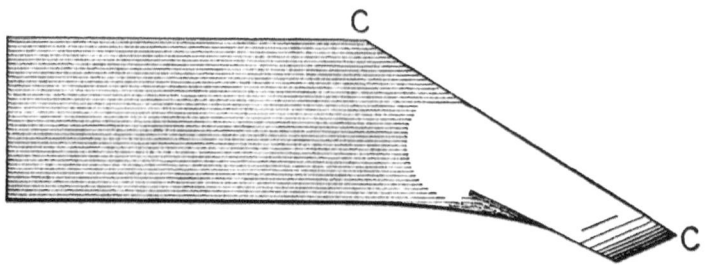

FIG. 337—SHOWING THE SHAPE GIVEN TO THE POINT.

Now I have described how I lay slip-lays. Solid bars or landsides I treat in the same way, for all the difference between the two is that there is no lip to make separately in the solid landsides. In laying the steel on the plow to mark the holes, I let it project over the bar one-eighth of an inch, which gives metal enough to dress a sharp corner in finishing.—*By* R. W. H.

MAKING A PLOWSHARE.

In making a plowshare, I first get out my steel like Fig. 338. Then I turn the point, as in Fig. 339. I do this so that when it is bent back the point will have the same angle as the edge of the share and will not be square or like a chisel, as it will if left straight and bent back square in the old way.

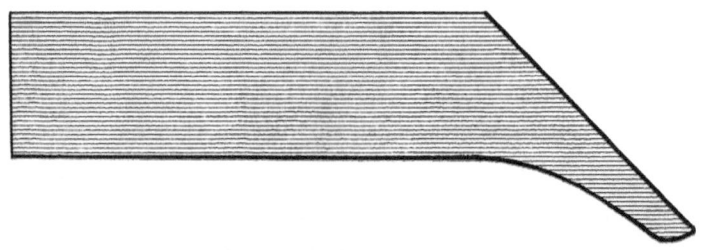

MAKING A PLOWSHARE.
FIG. 338—SHOWING HOW THE STEEL IS SHAPED AT FIRST.

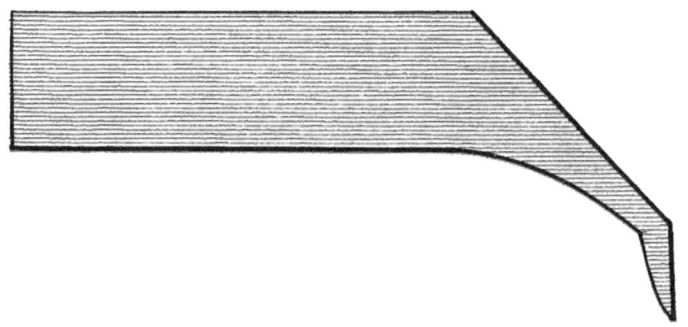

FIG. 339—SHOWING HOW THE POINT IS TURNED.

I next bend the point down and bring it back square with the shin. I next forge out the bar, and when about finished I take a light heat, and clamping it in the vise, draw a flange on the lip on the inside of the bar as in Fig. 340.

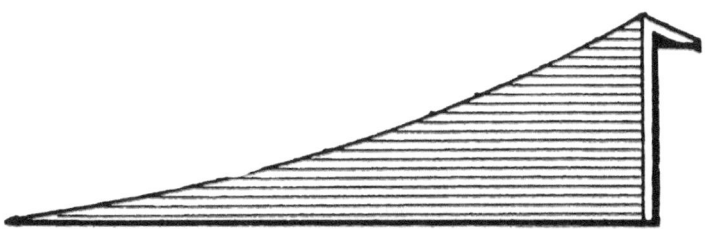

FIG. 340—SHOWING HOW THE FLANGE IS DRAWN.

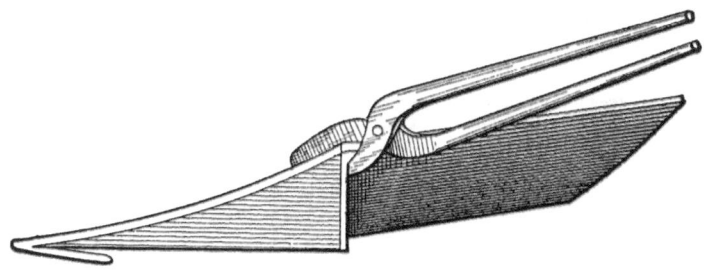

FIG. 341—SHOWING HOW THE BAR IS CLAMPED TO THE SHARE.

I now slip the point of the bar under the point of the share which is bent back, and with a pair of tongs I clamp the bar fast to the share by catching over the share and under the lip on the bar, as in Fig. 341, and in this way avoid trouble while welding.

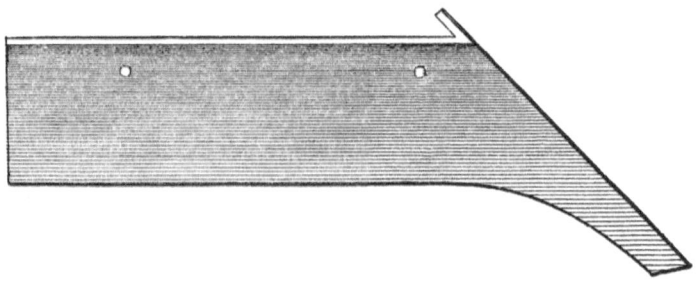

FIG. 342—SHOWING THE SHARE COMPLETED.

I begin at the point, and when near the top I turn the share over and with the pene of my hammer weld down the lip first, and then with the face of the hammer I strike on top of the share and never have failed to make a good weld in this way.—*By* L. H. O.

HOW TO SHARPEN A SLIP-SHEAR PLOW LAY.

Take iron ½ x ¾ inch square, thirty inches long, bend it in the center and bring the sides parallel with each other three-eighths of an inch apart, and weld the ends.

SHARPENING A SLIP-SHEAR PLOW LAY.
FIG. 343—THE PIECE WHICH PREVENTS THE LAY FROM SPRINGING.

This piece, shown in Fig. 343, is to keep the lay from springing up in the center. I then bolt this piece to the bottom of the lay with the three bolts taken out, or with new ones, as shown in Fig. 344, and then sharpen the edge of the lay from point to heel.

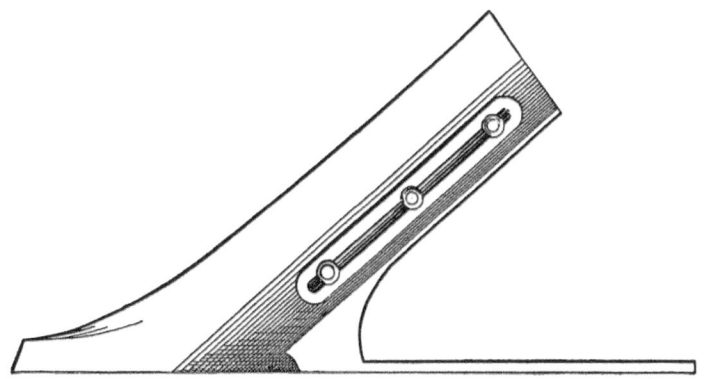

FIG. 344—SHOWING HOW THE PIECE IS BOLTED.

If there are no rocks where it is used I draw well back and very thin, and leave as few hammer marks as possible on top. I always set the edge from point to heel perfectly level with the landside on a level board or stone, not by just sighting with my eye. A level plow lay is bound to run well, and it will tickle the farmer all over when it runs well.—*By* G. W. P.

WELDING PLOW POINTS.

When making new points or welding old ones that have ripped, I turn the point bottom upwards, pour in a handful of wrought iron shavings along the seam, then a little borax on top of them, and lay the point in the fire just as it is. If care is taken to heat the bar a little the fastest, the shavings will come to a welding heat much sooner than the point, and will be like wax when the share and bar get to a welding heat.

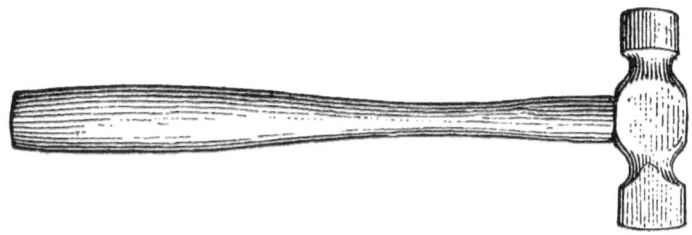

FIG. 345—HAMMER USED BY "J. W." IN WELDING PLOW POINTS.

Then with a light flat pened hammer, Fig. When the point is ready for tempering I lay it down and allow it to cool, then I heat the edge evenly from end to end and set it in the slack tub edge down, taking care that the edge touches the water evenly from end to end. By this means I make a point solid and unsprung.—*By* J. W.

HOW TO PUT NEW STEEL POINTS ON OLD PLOWS.

I have thought that someone would like to know how to make plow points last on rocky or clay land in Maine. The farmers use cast-iron plows mostly, and a new point don't last long. To help the poor farmer and myself just a bit, I new-steel old points by the following method: I use old carriage springs or old pieces of sled shoe steel, if I have them. First, take a piece of steel ¼ x 1 ¾ x 9 inches long for medium size plow, draw down one end thin, about one-eighth of an inch, and punch a five-sixteenths inch hole one inch from thin end, punch second hole four inches from first hole.

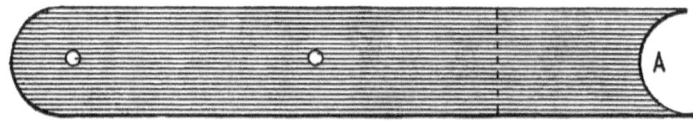

HOW TO PUT NEW STEEL POINTS ON OLD PLOWS.
FIG. 346—SHOWING HOW THE FIRST PIECE OF STEEL IS PREPARED.

Cut out the other end with a gouge-shaped chisel, as shown at *A*, in Fig. 346. Measure three inches from gouge cut end, and bend back, as shown at *B* Fig. 347.

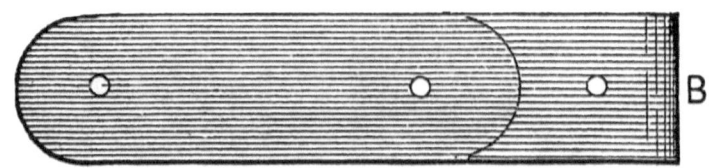

FIG. 347—SHOWING HOW FIRST PIECE OF STEEL IS BENT.

FIG. 348—SHOWING SECOND PIECE OF STEEL.

Cut off four and one-half inches of same size steel, draw down one end thin, say to one-eighth of an inch, and punch one-fourth inch hole in thick end as at *C*, Fig. 348.

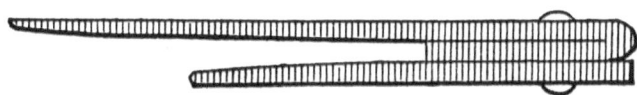

FIG. 349—SHOWING HOW THE TWO PIECES OF STEEL ARE BOLTED TOGETHER READY TO DRAW OUT.

Punch one-fourth inch hole in long piece, through the fold, rivet the two pieces together as in Fig. 349. Take welding heat on the parts that are riveted together and draw as in Fig. 350.

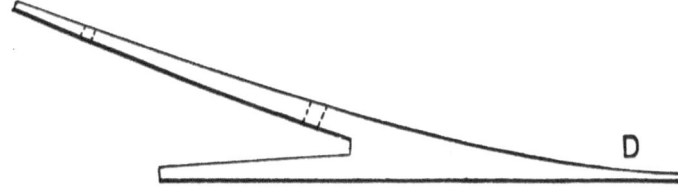

FIG. 350—SHOWING THE TWO PIECES OF STEEL SEEN IN FIG. 349 DRAWN TO A POINT.

Take the new point while hot and fit it to old one, work the first hole on old point drill, counter-sink and rivet on your new point; drill the second hole from top through cast iron and steel, countersink, be sure to cut the rivet plenty long enough to rivet, put it into the hole, and batter up the end just so the rivet won't fall out, then heat the point and rivet while the point is hot; fit new point to old one nicely while hot. If the old point did not carve the ground enough drop the end of the new point so as to be on a line with the bottom of plow as shown at E, Fig. 351.

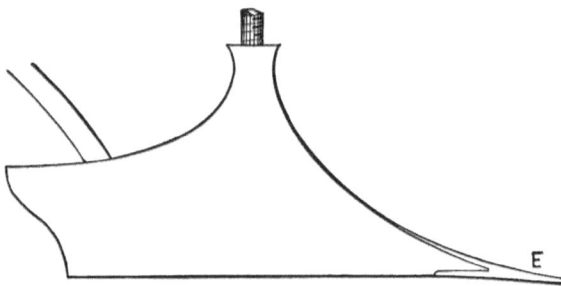

FIG. 351—SHOWING THE STEEL POINT ON THE PLOW.

Let the job cool before hardening the point. When done the new point should be from two and one-half to four inches long from the end of old point; for rocky ground they should not be as long as for clay. I charge from forty-five to seventy-five cents per point, according to size of plow, and the farmers say that a point fixed this way will do better work and outwear two new cast-iron points, which cost from sixty cents to one dollar each, thereby making quite a saving.—*By* geo. H. lambert.

POINTING PLOWS.

I first cut the point out of crucible steel one-fourth of an inch thick, as in Fig. 352. I draw the point to a thin edge as far back as the dotted line extends on that edge, then double the point back at *A* for right or left hand, as desired.

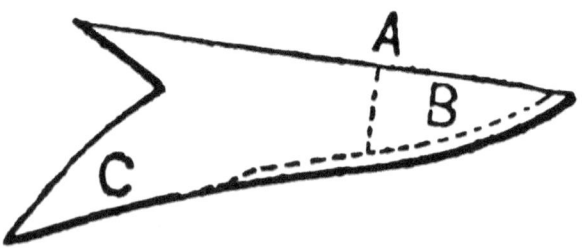

POINTING PLOWS, AS DONE BY "R. W. H."
FIG. 352—SHOWING THE POINT.

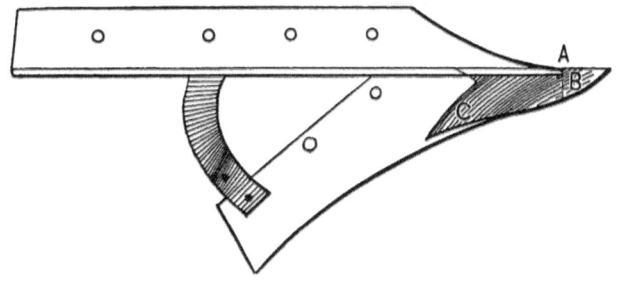

FIG. 353—SHOWING THE POINT APPLIED TO THE PLOW UNDERNEATH.

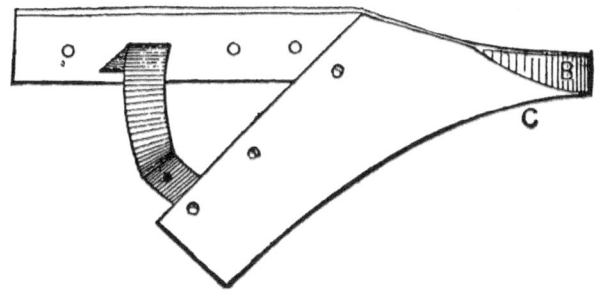

FIG. 354—SHOWING THE PART B ON TOP OF THE POINT.

I also thin out the point *C*. I forge down close, and after thinning the old point out, drive on as in Fig. 354, where the part *B* is shown on top of the point. In Fig. 353 the point is shown applied to the plow underneath. I next, with a large clean fire, weld on, commencing at the point, welding up the bar as far as the point extends, then having part *C* close to the share, weld up solid, draw out to make a full throat, and finish.

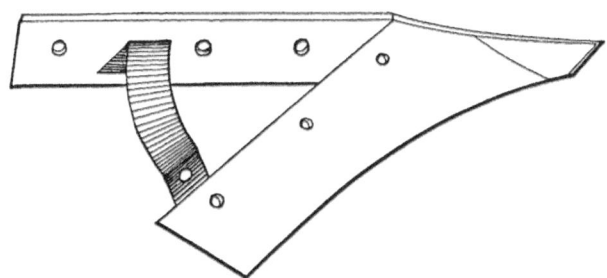

FIG. 335—SHOWING THE POINT COMPLETED.

Fig. 355 shows the point completed. This makes a very durable point, and always looks well if properly put on.—*By* R. W. H.

TEMPERING PLOW LAYS AND CULTIVATOR SHOVELS.

I have a recipe for tempering plow lays and cultivators which I think splendid. It is as follows: 1 lb. saltpetre, 1 lb. muriate ammonia, and 1 lb. prussiate of potash. Mix well, heat the steel to cherry red, and apply the powder lightly. When the oil is dry, cool in water. This leaves the steel very hard and also tough. The mixture is also good to case-harden iron and to make a heading tool.—*By* A. G. B.

SHARPENING LISTERS.

The prairies of the West are plowed, harrowed, and planted in corn with a single machine called a lister, and it is therefore probable that the method of sharpening it will be of interest to some readers.

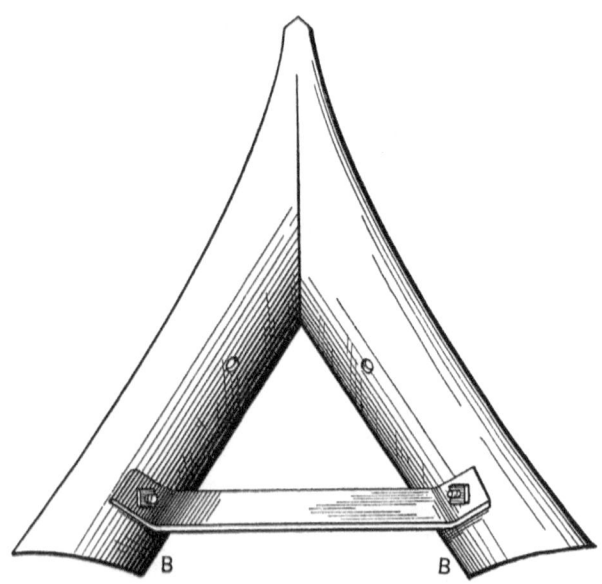

SHARPENING LISTERS.
FIG. 356—BOTTOM VIEW OF THE LISTER SHORE.

My way of doing the job is to take off the lister shore, Fig. 356, and after making the brace shown in Fig. 357, bolt this brace on the bottom side, so that the bolts in the back end of the lay will pull it a little.

Just here I will add, that as no two lister lays are alike, it is necessary to have a brace for each one, and it is advisable to mark each brace made, so that it can be used whenever the lay it fits comes back to the shop again. The next thing to do in the sharpening operation is to heat at the point and draw thin for three or four inches on one side. Then change to the other side and draw on that two or three inches more than you did on the first side taken in hand. Continue changing from one side to the other and testing it on a level surface.

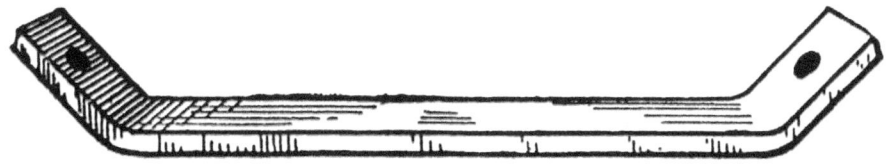

FIG. 357—THE BRACE.

It can be kept level and the point can be kept down, as the latter can be turned up easily by hammering on the bottom side.

Be sure to keep your lay level as you go on, and also keep it smooth on the top side. In some localities it must be polished to make it scour. Always let the lay get cold before you take the brace off, and then it can be put in place again without any trouble.—By G. W. Predmore.

NOTES ON HARROWS.

In a harrow I lately made I inserted lengths of one-half inch gas-pipe between the wooden bars, as sleeves on the rods or bolts, as shown in Fig. 358, so that all could be drawn up tight. It is quite a success. A good plan is to mortise in a light strap of iron, say 1 x ¼ inch, directly under the top strap, and bolt or rivet through, as at A in Fig. 359, all nuts on top, to keep the ground side smooth.

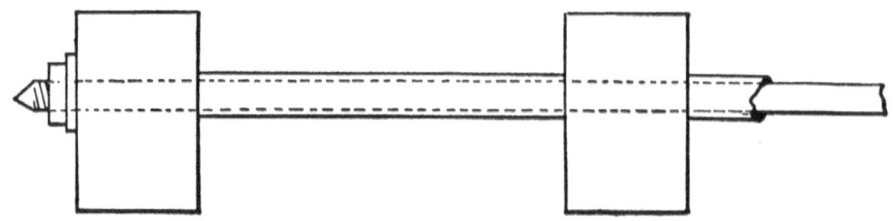

NOTES ON HARROWS. FIG, 358—SHOWING A METHOD OF UTILIZING GAS-PIPE IN MAKING A HARROW.

Fig. 360 is a top view of the parts shown in Fig. 359.

The narrow hinge shown in Fig. 359 is common, and can't be beaten. It allows each section to have a slight independent motion, can be unhooked at once by raising one part, and is easily folded over for cleaning.

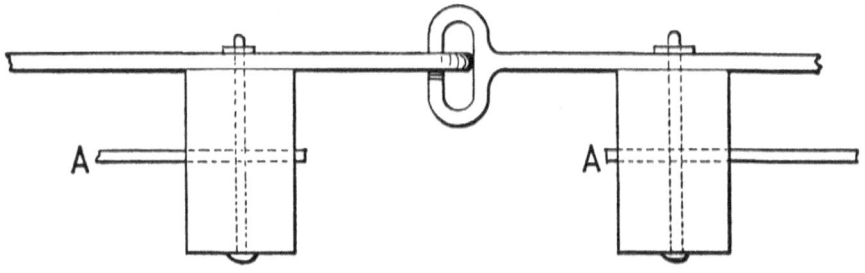

FIG. 359—SHOWING HOW THE LIGHT STRAP IS MORTISED IN AND HOW THE RIVETING AND BOLTING IS DONE.

I don't think it necessary to mortise the teeth holes out square, as is often done, round ones do well enough. A good size of tooth for general work is ½ x 5/8 inch (steel). This holds well in a five-eighths inch round hole. A round hole takes some driving, but I put a one-fourth inch bolt, or rivet ("wagon-box" head) and burr, back of each tooth. It may interest some to learn that in certain parts of Europe an iron tooth is inserted from below, having a shoulder to fit against the bottom of the bar and a thread and nut on top to hold it. It is pointed with steel, and when worn goes to the smith, like a plow, to be laid.

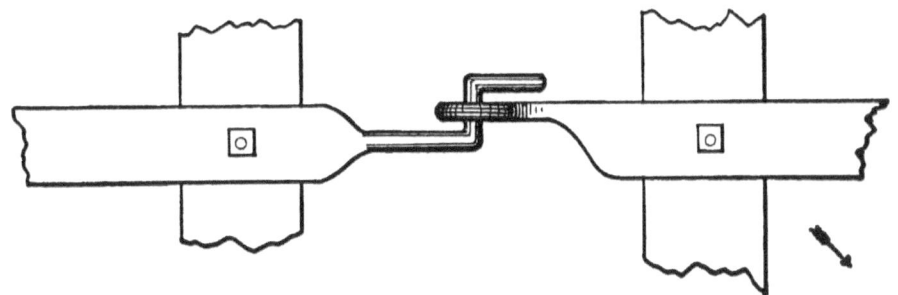

FIG. 360—TOP VIEW OF THE PARTS SHOWN IN FIG. 359.

I used a harrow of two-inch iron gas-pipe for many years, and the teeth (one-half inch square steel) held perfectly in round holes. The objection to it was that the cross-pieces being so low gathered up clods, etc.

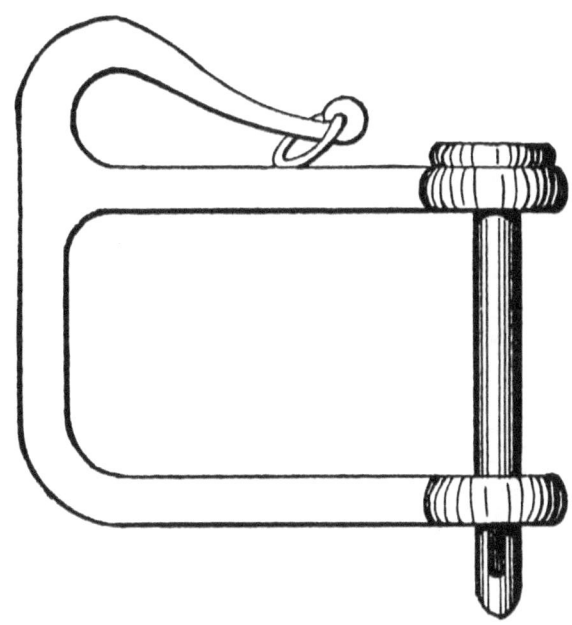

FIG. 361—SHOWING THE METHOD OF ATTACHING THE DOUBLETREE.

The best method of attaching the doubletrees is, I think, by a clevis combined with a safety-hook, as shown in Fig. 361.—*By* Will Tod.

MAKING A BOLT-HOLDER AND A PLOWSHARE.

A handy bolt-holder which I have occasion to use is made of two pieces of ⅜ x ¾ iron or steel, shape, put together as shown in Fig. 362. One piece is made with a flange to hinge into a slot, which is seen at *A* Fig. 362.

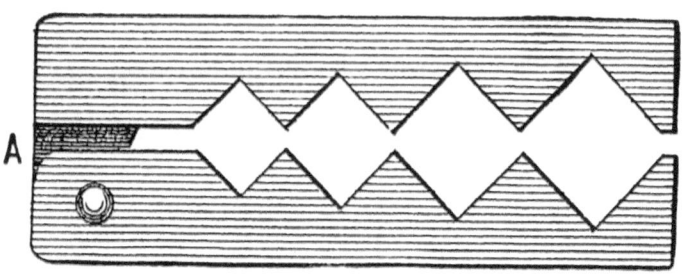

MAKING A BOLT-HOLDER AND PLOWSHARE
FIG. 362—SHOWING BOLT-HOLDER COMPLETE.

The notches are made different sizes so as to hold bolts of different sizes.

With this tool one can hold any plow bolt that has a countersunk head, and would be spoiled by the vise. The slot and rivet act as a hinge to take in large or small bolts.

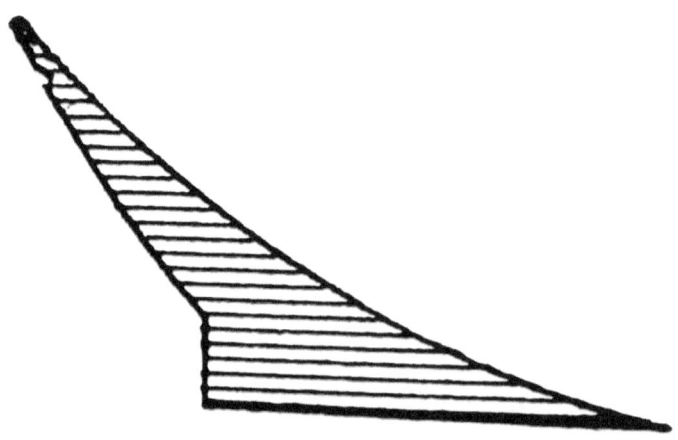

FIG. 363—SHOWING PIECE OF IRON AS USED BY "J. W. J." IN FITTING PLOWSHARE.

I have a great many shares to make for plows, and every smith knows how hard they are to fit. Instead of staving them, I take a piece of iron, as shown in Fig. 363, and weld on to the share, Fig. 364.

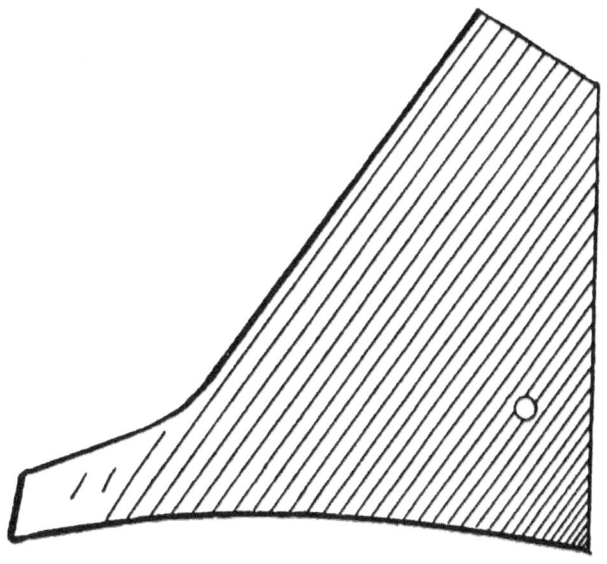

FIG. 364—FITTING PLOWSHARE.

After this is welded on it is an easy matter to fit it with a sharp chisel. I use five-sixteenths inch steel.—By J. W. J.

MAKING A GRUBBING HOE.

Plan 1.

The following is my plan of making a grubbing hoe or mattock. Take a piece of iron 2 x ½ inch, and about twelve inches long, cut it as at *A*, Fig. 365, bend open together and weld up solid to an inch and one-half of one end, then split open and put the steel in as at Fig. 366, then weld the other end for one and one-half inches together; this will leave about two inches not welded as at *B*; then take a heat not quite hot enough to weld at *B*, and forge as at *C*, Fig. 367, and *D*, leaving the eye, *C*, Fig. 367, closed.

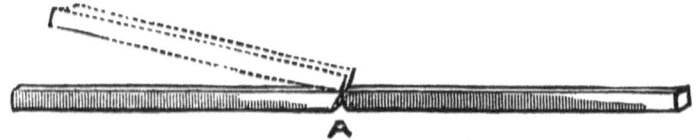

MAKING A GRUBBING HOE BY THE METHOD OF EPH. SHAW. FIG. 365—SHOWING THE IRON CUT FOR BENDING OPEN.

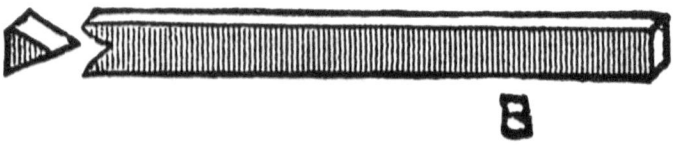

FIG. 366—SHOWING THE PIECE SPLIT TO INSERT THE STEEL.

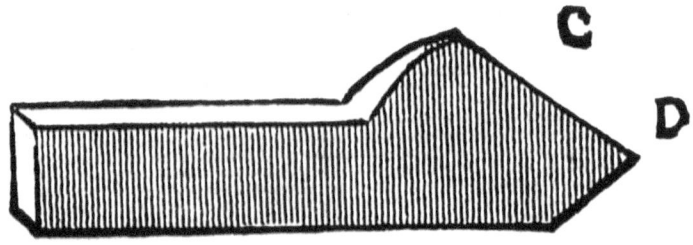

FIG. 367—SHOWING HOW THE IRON IS FORGED.

Then take a piece as before like Fig. 365, only ten inches long, and weld solid together throughout, and forge as Fig. 368, *E*.

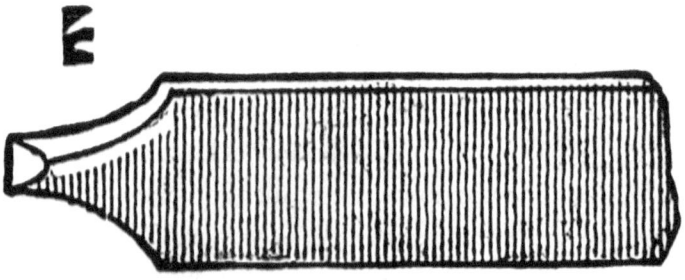

FIG. 368—SHOWING HOW THE FORGING IS DONE AT G, ON THE OTHER PIECE.

Then take a good welding heat on both pieces and weld as at *F*, Fig, 369, with steel in as *G*, then forge and finish up as at Figs. 370 and 371.

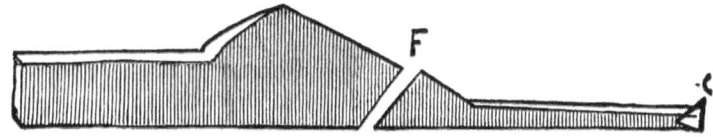

FIG. 369—SHOWING HOW THE TWO PIECES ARE WELDED.

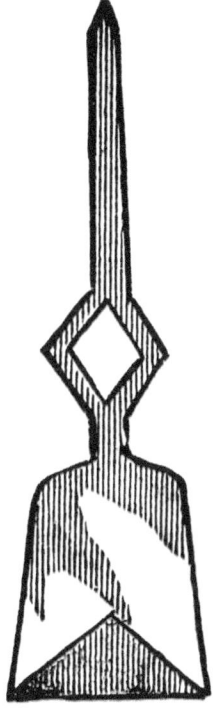

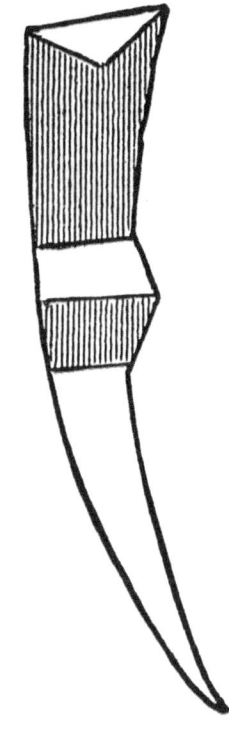

FIG. 370—SHOWING THE FINISHED MATTOCK.

FIG. 371—ANOTHER VIEW OF THE FINISHED MATTOCK.

This makes a good strong mattock, and is the only way I know of to make one, unless it is to weld two pieces together and form the eye, and then twist for the hoe end.—*By* Eph. Shaw.

MAKING A GRUBBING HOE.

Plan 2.

To make a grubbing hoe, take iron 3 x ½ inch, cut as shown in Fig. 372, draw out the ends, bend at *A* to a right angle, bring *B B* together, as shown in Fig. 373, and then weld.

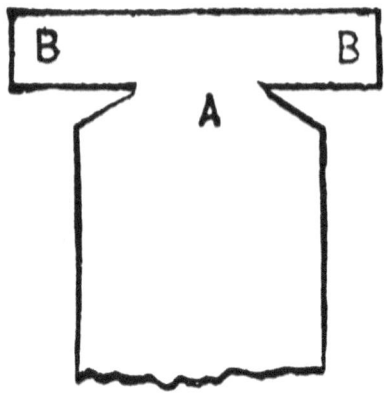

MAKING A GRUBBING HOE BY "SOUTHERN BLACKSMITH'S" METHOD. FIG. 372—THE IRON CUT AND DRAWN.

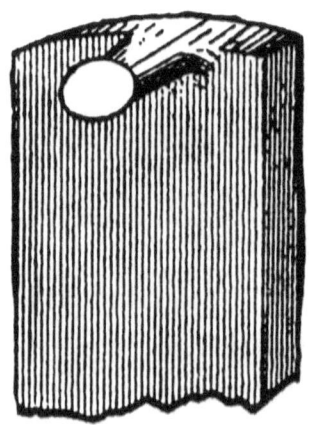

FIG. 373—THE IRON BENT TO SHAPE.

This is an easy job, and the result is a good hoe.—*By* Southern Blacksmith.

FORGING A GARDEN RAKE.

The question is often asked, Can forks be made successfully of cast-steel? I can always forge better with cast-steel than with any other metal. I have made a garden rake of cast-steel and it was a good, substantial job. It is done as follows: Take a piece of steel one-fourth or five-sixteenths of an inch thick, lay off the center as in Fig. 374, then punch a hole about as far from the center as is necessary to give stock enough to turn at a right angle for the stem to go in the handle. Then cut out with a sharp chisel as marked in the dotted lines A. Then lay off the teeth B B.

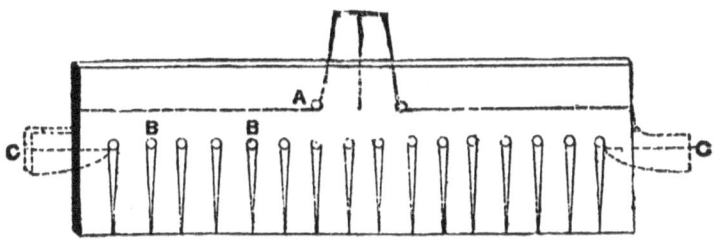

FORGING A GARDEN RAKE.
FIG. 374—"CONSTANT READER'S" PLAN.

Punch or drill holes and cut out. The end pieces C C can be turned out straight and drawn out well. After the holes and the pieces are cut out, to separate the teeth turn each tooth (one at a time) at right angles, and draw out to the desired size. Then straighten it back to its place, and so proceed until all the teeth are drawn. By using a tool with holes in it to suit the tooth you can give it a good finish on the anvil.—*By* Constant Reader.

MAKING A DOUBLE SHOVEL PLOW.

I will describe my method of making a double shovel plow: I first make the irons. The shovels should be five inches wide, twelve inches long, and cut to a diamond or a shovel shape, as the customer desires. After drawing, bend a true arc from point to top, on a circle of twenty-two inches in diameter. The plow will then, as it wears away, retain the same position it had when new. I make the faces of plows to suit customers.

Some prefer them flat, others want them oval, and some want a ridge up the middle. In the latter style a cross section of the plow would look as in Fig. 375.

MAKING A DOUBLE SHOVEL PLOW.
FIG. 375—CROSS SECTIONAL VIEW OF A RIDGE-FACED PLOW.

The next thing to be done is to make four brace rods, two one-half inch, and two ½ x 15 inches. There should be ten inches between the center of the eye and the nail hole. I cut threads on the ends of the one-half inch rods, and punch nail hole in end of three-eighths inch rods. I then take three bolts, two one-half inch, one ⅜ x 3 inches long; one of the one-half inch bolts should be eight inches, and the other seven and one-half inches long.

The clevis comes next. I take two pieces of one-half inch round iron, ten inches long, flatten one end of each piece, punch three-eighths inch hole, lay the flat ends together, weld the round ends and bend to shape, when the clevis will look as in Fig. 376.

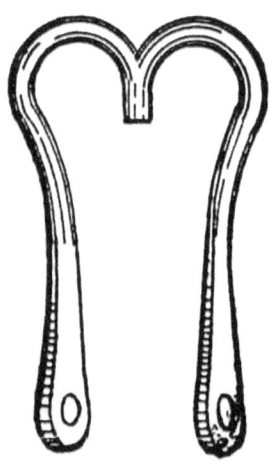

FIG. 376—THE CLEVIS.

I bore a hole in the end of the beam to admit the point of the clevis. This keeps it firmly in place. The irons are then finished. In beginning on the woodwork, I take two uprights or standards, one three and one-half feet long, and three and one-fourth inches in the widest part, and two inches thick; the other is of the same dimensions, except that it is somewhat shorter at the top end. I make these as in Fig. 377.

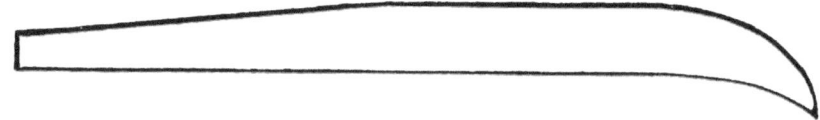

FIG. 377—SHOWING THE STANDARD.

The beam is made four feet, three inches long, three and one-half inches wide in the widest part, and two and one-half inches at the point. I bore three holes in the wide part, two holes being one-half inch, and one three-eighths inch, and each being twelve inches from center to center, as shown in Fig. 378.

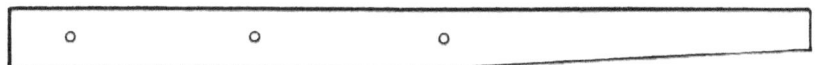

FIG. 378—THE BEAM.

After making the handles I fit the shovels to the uprights, and then take two strips of plank 1x2 inches, one strip fifteen inches, the other sixteen inches long, and nail blocks on one end, as in Fig. 379, and nail them down to the floor so that the ends of beam will rest in them, as shown in the illustration, resting the front end on the short piece. I lay the beam on, take an upright, stand it up by the side as in Fig. 379, and adjust until the shovel stands on floor to suit me.

I let the shovels stand rather flat on the floor. They will run better when sharp, but will not wear as long without sharpening. I put the pencil through the hole in beam, and mark the place where the hole is to be bored in the upright pieces. I bore brace holes and take a seven and one-half inch bolt, and put it through the long upright from left to right.

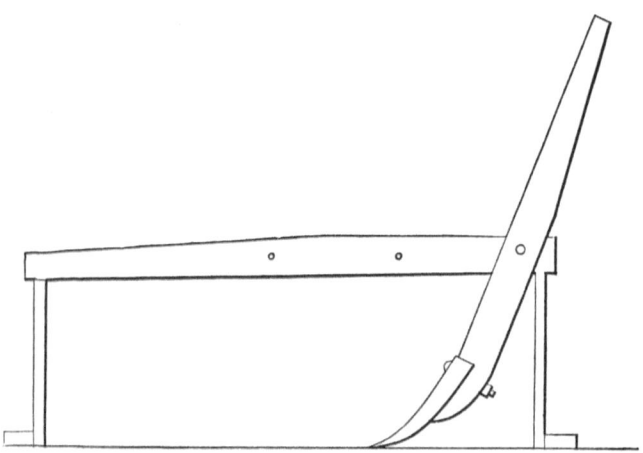

FIG. 379—SHOWING THE METHOD OF ADJUSTING THE POSITION OF THE STANDARD.

The block put on should be 3 x 3 inches, and two and one-half inches long. Some use iron, but wood is as good and makes the plow lighter. I put the bolts through the beam and put on a three-eighths inch brace and a nut. I put an eight-inch bolt through the short upright block and beam from right to left, and put a one-half inch brace through the hole in the upright, as shown in Fig. 379. I put the eye of the one-half inch brace on the bolt and also use a three-eighths inch brace.

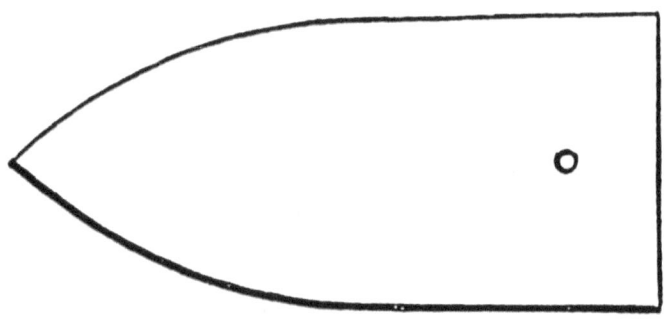

FIG. 380—THE SHOVEL.

I use a one-half inch brace in the front upright, and also one bolt. I screw the nuts up, bend three-eighths inch brace down under the beam and against uprights, rivet the end to the upright, and the plows will then stay well apart.

In placing the handle on the beam I ascertain the height wanted on the upright, bore an inch hole through the same, and fit the rung in so that it projects three inches to the left of the upright.

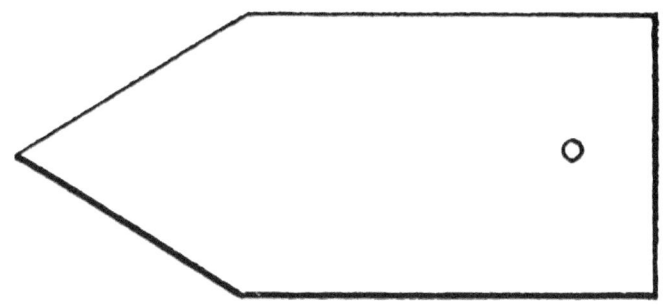

FIG. 381—THE DIAMOND-SHAPED SHOVEL.

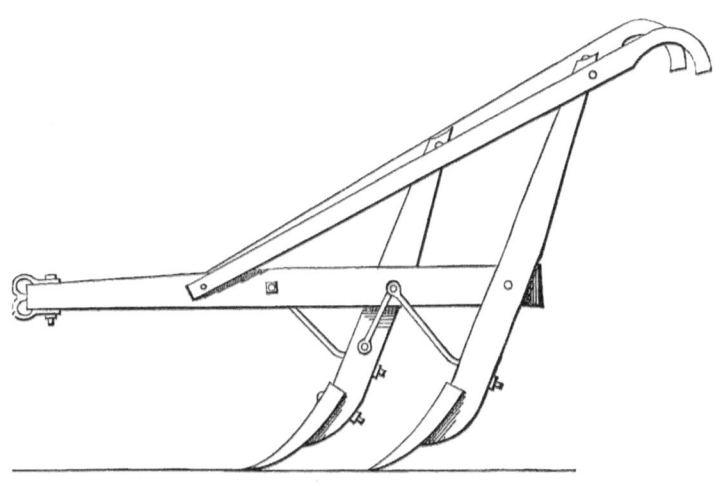

FIG. 382—THE FINISHED PLOW.

The right handle is put on so as to come on the outside of the front upright. I notch the upright to fit the handle and bolt them together with a five-sixteenths inch bolt. Then the clevis is put on and the plow is ready for painting. Fig. 380 represents the ordinary shovel, and Fig. 381 shows the diamond-shaped shovel.

The finished plow is shown in Fig. 382.—*By* C. Jake.

POINTING CULTIVATOR SHOVELS.

Plan 1.

We that labor for farmers must know how to do work on farm implements and machines, and so there may be many who would like to know a good way for pointing cultivator shovels. My plan is as follows:

I take a piece of spring steel about six inches long and one and one-half inches wide, and draw it out from the center toward each end to the shape shown in Fig. 383. I then draw out the straight side *A* to a thin edge and cut through the dotted line nearly to the point.

POINTING CULTIVATOR SHOVELS BY THE METHOD OF "A. M. B."
FIG. 383—SHOWING THE SHAPE TO WHICH THE STEEL IS DRAWN.

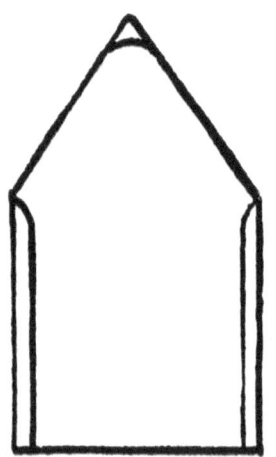 

FIG. 384—SHOWING THE OLD CULTIVATOR SHOVEL.

FIG. 385—SHOWING THE STEEL AFTER IT HAS BEEN CUT INTO AND DOUBLED AROUND.

I next double around as in Fig. 385 and take a light heat to hold it solid. I then take the old cultivator shovel, as illustrated in Fig. 384, straighten it out flat, lay the point on the back side, take a couple of good welding heats, and finish up to shape as in Fig. 385, making virtually a new shovel.—*By* A. M. B.

POINTING CULTIVATOR SHOVELS.

Plan 2.

I will describe my way of pointing cultivator shovels, and I think it is the best and most economical I ever tried. Take a piece of one-quarter inch plow steel two and one-half inches wide, cut it as shown in Fig. 386, and forge it into the shape shown in Fig. 387.

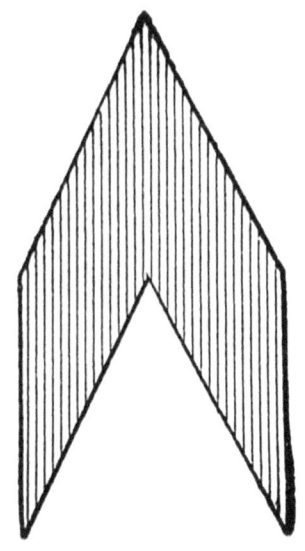
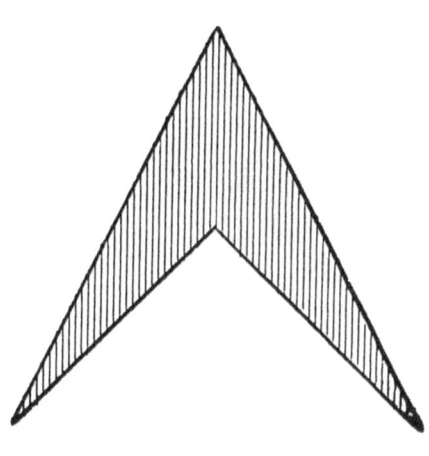

POINTING CULTIVATOR SHOVELS. FIG. 386—SHOWING HOW THE STEEL IS CUT.

FIG. 387—SHOWING THE POINT AFTER THE FORGING.

Then you will have a point with just the right amount of steel in the right place and with none wasted. I always mark the whole bar with a dot punch before I commence cutting the points.—*By* W. L. S.

POINTING CULTIVATOR SHOVELS.

Plan 3.

I cut out my points in one piece as shown in Fig. 388, using good crucible steel. Then I take the shovel to be pointed and draw its point, then half way up each edge I also draw the points shown in Fig. 388, where they lap on to the shovel edge, and cut them to fit under the point of the shovel to be pointed, as shown in Fig. 389.

POINTING CULTIVATOR SHOVELS. FIG. 388—SHOWING THE POINT.

FIG. 389—THE POINT ADJUSTED READY FOR WELDING.

In getting the point in place clamp it with a pair of tongs on the right side of the shovel, put in the fire on the left side, with the face of the shovel up; have an open, clear fire ready, put on borax, and let the heat come up slowly, so as not to burn the point underneath.

A fan blower is better for this purpose, as you can regulate it to any degree of blast. After welding each point, take a light heat and tap the thin edge of the shovel down snug upon the point and lay it flat in the fire, face up; then take a wide heat (with plenty of borax) over the entire point and weld down solid with quick blows and with a hammer a little rounded on the face. Draw the edges out thin and hold the piece on the anvil so as to bevel from the bottom, leaving the top of the shovel level, as in Fig. 390.

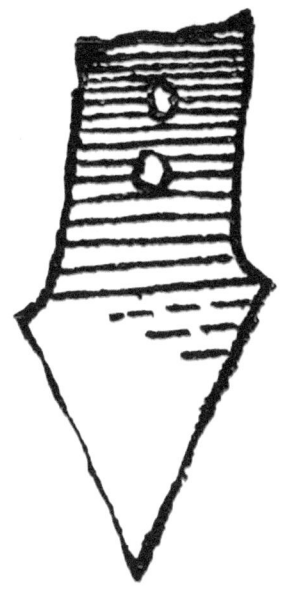

FIG. 390—THE FINISHED SHOVEL.

After finishing, heat to a cherry red all over, and plunge it in water edgeways and perpendicular, holding it still in one place till cool, then grind the scale off of the face and polish it on an emery belt. Shovels pointed in this way are almost as good as new, and will scour and give perfect satisfaction.—*By* R. W. H.

END OF VOLUME III

www.ingramcontent.com/pod-product-compliance
Lightning Source LLC
Chambersburg PA
CBHW020419010526
44118CB00010B/323